Quantum Mechanics

Quantum Mechanics
Problems and Solutions

K. Kong Wan

JENNY STANFORD
PUBLISHING

Published by

Jenny Stanford Publishing Pte. Ltd.
Level 34, Centennial Tower
3 Temasek Avenue
Singapore 039190

Email: editorial@jennystanford.com
Web: www.jennystanford.com

British Library Cataloguing-in-Publication Data
A catalogue record for this book is available from the British Library.

Quantum Mechanics: Problems and Solutions

ISBN 978-981-4800-72-3 (Paperback)
ISBN 978-0-429-29647-5 (eBook)

To my beautiful granddaughter Orly Rose,
whose new arrival brings
infinite joy and jubilation.

Contents

Preface

This is a solutions manual to accompany the book **Quantum Mechanics: A Fundamental Approach** by the author published in 2019 by Jenny Stanford Publishing, Singapore. It provides detailed solutions to all the questions listed at the end of each chapter of the book, except for the introductory Chapters 1 and 2. These questions are reproduced here chapter by chapter, followed by their solutions, which are labelled to correspond to the questions. For example, SQ3(1) is the solution to question Q3(1), which is the first question listed in Exercises and Problems at the end of Chapter 3 of the book.

The solutions presented make full use of the materials in the book. All the theorems, definitions, examples, comments, properties, postulates and equations in the book are referred to by their chapter or section numbers. For instance, Theorem 13.3.2(2) refers to the second theorem in section 13.3.2, Eq. (4.18) refers to equation (4.18) in Chapter 4, P15.1(5) refers to property (5) in section 15.1, and C28.2(3) refers to comment (3) in section 28.2. Equation labelling in terms of (∗), (∗∗), (∗∗∗) and (∗∗∗∗) is introduced here in some questions when they are needed for reference later in their solutions.

K. Kong Wan
St Andrews
Scotland

Chapter 1

Structure of Physical Theories

This introductory chapter sets out a general structure of physical theories which is applicable to both classical and quantum mechanics. We start with measurable properties of a given physical system, be it classical or quantum. These properties are called *observables*. We then introduce a definition of the state of a physical system in terms of measured values of a sufficiently large set of observables. A theory to describe the system should consists of four basic components:

1. *Basic mathematical framework* This comprises a set of elements endowed with some specific mathematical structure and properties. In mathematics such a set is generally known as a **space**.

2. *Description of states* States are described by elements of the space in the chosen mathematical framework. For this reason the space is called the **state space** of the system.

3. *Description of observables* Observables are to be described by quantities defined on the state space. The description should yield all possible values of observables. The relationship between observables and states should be explicitly stated. The following two cases are of particular interest:

Quantum Mechanics: Problems and Solutions
K. Kong Wan

ISBN 978-981-4800-72-3 (Paperback), 978-0-429-29647-5 (eBook)
www.jennystanford.com

(1) For a deterministic theory like classical mechanics a state should determine the values of all observables.

(2) For a probabilistic theory like quantum mechanics a state should determine the probability distribution of the values of all observables.

4. *Description of time evolution (dynamics).*

Chapter 2

Classical Systems

This chapter sets out some general physical properties of classical systems which are divided into discrete and continuous:

1. *Discrete systems* These are systems of discrete point particles. The specific structure of classical mechanics is presented with position, linear and angular momenta serving as basic observables.

2. *Continuous systems* These systems are illustrated by a vibrating string. Continuous systems have different kinds of properties and observables, e.g., wave properties. In particular we have discussed:

(1) Description of states by solutions of the classical wave equation.
(2) The concept of eigenfunctions with orthonormality property and their superposition and interference.
(3) The concept of a complete set of states.

These discussions are given specifically to provide an intuition to help a better understanding of similar properties of quantum systems.

Quantum Mechanics: Problems and Solutions
K. Kong Wan

ISBN 978-981-4800-72-3 (Paperback), 978-0-429-29647-5 (eBook)
www.jennystanford.com

Chapter 3

Probability Theory for Discrete Variables

Q3(1) Prove Theorem 3.4(1).

SQ3(1) Theorem 3.4(1) can be proved using properties PM3.4(1), PM3.4(2) and PM3.4(3) in Definition 3.4(2).

(1) To prove Eq. (3.25) let E be any event. Then $E \cup \emptyset = E$ and $E \cap \emptyset = \emptyset$. By PM3.4(3) we have,

$$\left.\begin{array}{l}\mathcal{M}_p(E \cup \emptyset) = \mathcal{M}_p(E) \\ \mathcal{M}_p(E \cup \emptyset) = \mathcal{M}_p(E) + \mathcal{M}_p(\emptyset)\end{array}\right\} \Rightarrow \mathcal{M}_p(\emptyset) = 0.$$

(2) To prove Eq. (3.26) we start with $\mathcal{M}_p(S_{am}) = 1$ by PM3.4(2). Since $E \cap E^c = \emptyset$ and $E \cup E^c = S_{am}$ we have

$$\begin{aligned}\mathcal{M}_p(S_{am}) &= \mathcal{M}_p(E \cup E^c) = \mathcal{M}_p(E) + \mathcal{M}_p(E^c) = 1 \\ &\Rightarrow \quad \mathcal{M}_p(E^c) = 1 - \mathcal{M}_p(E).\end{aligned}$$

(3) To prove Eq. (3.27) we first observe that $E_1 \subset E_2 \Rightarrow E_2 = (E_2 - E_1) \cup E_1$. Since $(E_2 - E_1)$ and E_1 are disjoint, i.e., $(E_2 - E_1) \cap E_1 = \emptyset$, we have

$$\begin{aligned}\mathcal{M}_p(E_2) &= \mathcal{M}_p\Big((E_2 - E_1) \cup E_1\Big) \\ &= \mathcal{M}_p(E_2 - E_1) + \mathcal{M}_p(E_1) \geq \mathcal{M}_p(E_1).\end{aligned}$$

Quantum Mechanics: Problems and Solutions
K. Kong Wan

ISBN 978-981-4800-72-3 (Paperback), 978-0-429-29647-5 (eBook)
www.jennystanford.com

(4) The proof of Eq. (3.28) is based on the decomposition, obvious from the Venn diagram,

$$E_1 = (E_1 - E_2) \cup (E_1 \cap E_2),$$

where $(E_1 - E_2)$ and $(E_1 \cap E_2)$ are disjoint. Then PM3.4(3) implies

$$\mathcal{M}_p(E_1) = \mathcal{M}_p(E_1 - E_2) + \mathcal{M}_p(E_1 \cap E_2),$$

which is Eq. (3.28).
Two cases are noteworthy:

(a) E_1 and E_2 are disjoint, i.e., $E_1 \cap E_2 = \emptyset$. Then $E_1 - E_2 = E_1$, we have

$$\mathcal{M}_p(E_1 \cap E_2) = 0 \quad \text{and} \quad \mathcal{M}_p(E_1 - E_2) = \mathcal{M}_p(E_1).$$

Equation (3.28) is again satisfied.

(b) E_1 is a subset of E_2. Then

$$E_1 \cap E_2 = E_1 \quad \text{and} \quad E_1 - E_2 = \emptyset.$$

Equation (3.28) is again satisfied in these two cases.

(5) To prove Eq. (3.29) we first note that if E_1 and E_2 are disjoint the equation is obviously true, i.e.,

$$\mathcal{M}_p(E_1 \cup E_2) = \mathcal{M}_p(E_1) + \mathcal{M}_p(E_2)$$

by PM3.4(3). If E_1 and E_2 are not disjoint, then the Venn diagram tells us that

$$E_1 \cup E_2 = (E_1 - E_2) \cup E_2,$$

where $(E_1 - E_2)$ and E_2 are disjoint. We have, by PM3.4(3),

$$\mathcal{M}_p(E_1 \cup E_2) = \mathcal{M}_p(E_1 - E_2) + \mathcal{M}_p(E_2).$$

Using Eq. (3.28) we immediately get

$$\mathcal{M}_p(E_1 \cup E_2) = \mathcal{M}_p(E_1) + \mathcal{M}_p(E_2) - \mathcal{M}_p(E_1 \cap E_2),$$

which is Eq. (3.29).
We have assumed E_2 is a subset of E_1 in the above proof. When E_1 is a subset of E_2 we can similarly establish the result.

Q3(2) Prove Theorem 3.5(1).

SQ3(2) Equation (3.36) is proved as follows:

$$\begin{aligned}\text{Var}\,(\wp^A) &= \sum_{\ell=1}^{n} \left(a_\ell - \mathcal{E}(\wp^A)\right)^2 \wp^A(a_\ell) \\ &= \sum_{\ell=1}^{n} \left(a_\ell^2 - 2a_\ell\,\mathcal{E}(\wp^A) + \mathcal{E}(\wp^A)^2\right)\wp^A(a_\ell) \\ &= \left(\sum_{\ell=1}^{n} a_\ell^2 \wp^A(a_\ell)\right) - \mathcal{E}(\wp^A)^2,\end{aligned}$$

using the fact that

$$\sum_{\ell=1}^{n} \left(-2a_\ell\,\mathcal{E}(\wp^A)\right)\wp^A(a_\ell) = -2\mathcal{E}(\wp^A)^2 \text{ and } \sum_{\ell=1}^{n} \wp^A(a_\ell) = 1.$$

Q3(3) What is the value $\mathcal{F}(a_4) - \mathcal{F}(a_3)$ of the probability distribution function in Eq. (3.38)?

SQ3(3) The value $\mathcal{F}(a_4) - \mathcal{F}(a_3)$ is equal to $\wp(a_4)$.

Q3(4) In an experiment of tossing a fair die a number from 1 to 6 will be obtained with equal probabilities.

(a) Write down the probability mass function $\wp$ and evaluate the expectation value and the uncertainty.

(b) Write down the corresponding probability distribution function $\mathcal{F}(\tau)$ and sketch a plot of $\mathcal{F}(\tau)$ versus τ. What are the values $\mathcal{F}(\tau)$ at $\tau = 0.9, 1, 2.5, 6$ and 6.1?

SQ3(4)(a) For a fair die every number is equally likely to appear in a toss. The probability mass function is a function $\wp$ on the sample space $S_{am} := \{a_1 = 1, a_2 = 2, a_3 = 3, a_4 = 4, a_5 = 5, a_6 = 6\}$ defined by $\wp(a_\ell) = 1/6$ for all $a_\ell \in S_{am}$. The expectation value is

$$1 \times \frac{1}{6} + 2 \times \frac{1}{6} + 3 \times \frac{1}{6} + 4 \times \frac{1}{6} + 5 \times \frac{1}{6} + 6 \times \frac{1}{6} = 3.5.$$

The variance is given by Theorem 3.5(1) to be

$$\left(1 \times \frac{1}{6} + 4 \times \frac{1}{6} + 9 \times \frac{1}{6} + 16 \times \frac{1}{6} + 25 \times \frac{1}{6} + 36 \times \frac{1}{6}\right) - 3.5^2$$

$$\approx 2.9.$$

The uncertainty $= \sqrt{\text{variance}}$ with an approximate value $\sqrt{2.9}$.

SQ3(4)(b) The probability distribution function $\mathcal{F}(\tau)$ is piecewise-constant with discontinuous steps occurring at $\tau = 1, 2, 3, 4, 5, 6$. Explicitly $\mathcal{F}(\tau)$ is related to $\wp(a_\ell)$ by

$$\mathcal{F}(\tau) = \begin{cases} 0, & \tau < 1 \\ 1/6, & 1 \le \tau < 2 \\ 2/6, & 2 \le \tau < 3 \\ 3/6, & 3 \le \tau < 4 \\ 4/6, & 4 \le \tau < 5 \\ 5/6, & 5 \le \tau < 6 \\ 1, & 6 \le \tau \end{cases}.$$

The function $\mathcal{F}(\tau)$ is continuous from the right for all τ. A plot of $\mathcal{F}(\tau)$ against τ is given below:

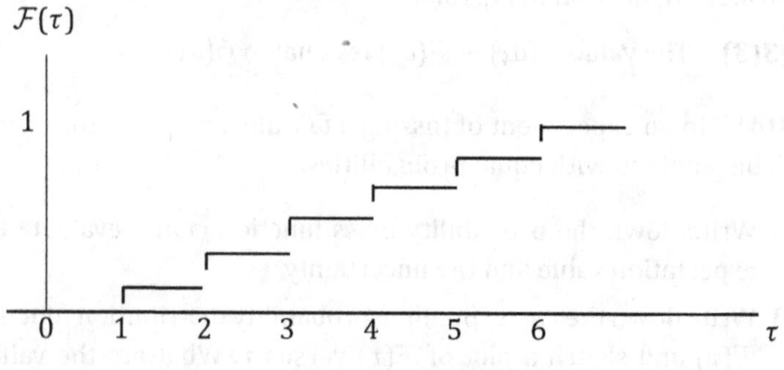

We have $\mathcal{F}(0.9) = 0$, $\mathcal{F}(1) = 1/6$, $\mathcal{F}(2.5) = 2/6$, $\mathcal{F}(6) = 1$, $\mathcal{F}(6.1) = 1$.

Chapter 4

Probability Theory for Continuous Variables

Q4(1) For the characteristic function χ_Λ show that the inverse image of every Borel set is a Borel set.

SQ4(1) Let Λ' be a Borel set in the codomain of the characteristic function $\chi_\Lambda(\tau)$ which is be taken as $I\!R$. Then:

(1) If Λ' does not contain the value 0 or 1 the inverse image $\chi_\Lambda^{-1}(\Lambda') = \emptyset$.

(2) If Λ' contain the value 0 but not 1 the inverse image $\chi_\Lambda^{-1}(\Lambda') = \Lambda^c$, the complement of Λ.

(3) If Λ' contain the value 1 but not 0 the inverse image $\chi_\Lambda^{-1}(\Lambda') = \Lambda$.

(4) If Λ' contain both the value 0 and 1 the inverse image $\chi_\Lambda^{-1}(\Lambda') = I\!R$.

All the inverse images are Borel sets $\Rightarrow \chi_\Lambda(\tau)$ is a Borel function.

Q4(2) Prove Eq. (4.12).

SQ4(2) Use the additive and the defining properties of Lebesgue measure, i.e., the Lebesgue measure of a singleton set is zero and that the measure of a half-open interval $(\tau_1, \tau_2]$ is equal to $\tau_2 - \tau_1$.

Quantum Mechanics: Problems and Solutions
K. Kong Wan

ISBN 978-981-4800-72-3 (Paperback), 978-0-429-29647-5 (eBook)
www.jennystanford.com

Since $[\tau_1, \tau_2] = \{\tau_1\} \cup (\tau_1, \tau_2]$ and $\{\tau_1\} \cap (\tau_1, \tau_2] = \emptyset$ we get

$$\begin{aligned}\mathcal{M}_l\big([\tau_1, \tau_2]\big) &= \mathcal{M}_l\big(\{\tau_1\} \cup (\tau_1, \tau_2]\big) \\ &= \mathcal{M}_l\big(\{\tau_1\}\big) + \mathcal{M}_l\big((\tau_1, \tau_2]\big) = \mathcal{M}_l\big((\tau_1, \tau_2]\big) \\ \Rightarrow \quad & \mathcal{M}_l\big([\tau_1, \tau_2]\big) = \tau_2 - \tau_1. \qquad (*)\end{aligned}$$

We can prove two similar cases:

(1) To prove $\mathcal{M}_l\big([\tau_1, \tau_2)\big) = \tau_2 - \tau_1$ we have

$$\begin{aligned}\mathcal{M}_l\big([\tau_1, \tau_2]\big) &= \mathcal{M}_l\big([\tau_1, \tau_2) \cup \{\tau_2\}\big) \\ &= \mathcal{M}_l\big([\tau_1, \tau_2)\big) + \mathcal{M}_l\big(\{\tau_2\}\big) = \mathcal{M}_l\big([\tau_1, \tau_2)\big) \\ \Rightarrow \quad & \mathcal{M}_l\big([\tau_1, \tau_2)\big) = \tau_2 - \tau_1 \quad \text{by Eq. } (*) \text{ above.}\end{aligned}$$

(2) To prove $\mathcal{M}_l\big((\tau_1, \tau_2)\big) = \tau_2 - \tau_1$ we have

$$\begin{aligned}\mathcal{M}_l\big((\tau_1, \tau_2]\big) &= \mathcal{M}_l\big((\tau_1, \tau_2) \cup \{\tau_2\}\big) \\ &= \mathcal{M}_l\big((\tau_1, \tau_2)\big) + \mathcal{M}_l\big(\{\tau_2\}\big) = \mathcal{M}_l\big((\tau_1, \tau_2)\big) \\ \Rightarrow \quad & \mathcal{M}_l\big((\tau_1, \tau_2)\big) = \tau_2 - \tau_1 \quad \text{by Eq. } (*) \text{ above.}\end{aligned}$$

Q4(3) Show that the Lebesgue measure of the set of irrational numbers in $[a, b]$ is equal to $b - a$.

SQ4(3) Let the set of rational and irrational numbers in $[\tau_1, \tau_2]$ be denoted by $[\tau_1, \tau_2]_{ra}$ and $[\tau_1, \tau_2]_{ir}$ respectively. Then we have $[\tau_1, \tau_2] = [\tau_1, \tau_2]_{ra} \cup [\tau_1, \tau_2]_{ir}$ and $[\tau_1, \tau_2]_{ra} \cap [\tau_1, \tau_2]_{ir} = \emptyset$. It follows that

$$\begin{aligned}\mathcal{M}_l\big([\tau_1, \tau_2]\big) &= \mathcal{M}_l\big([\tau_1, \tau_2]_{ra} \cup [\tau_1, \tau_2]_{ir}\big) \\ &= \mathcal{M}_l\big([\tau_1, \tau_2]_{ra}\big) + \mathcal{M}_l\big([\tau_1, \tau_2]_{ir}\big) = \mathcal{M}_l\big([\tau_1, \tau_2]_{ir}\big),\end{aligned}$$

since $[\tau_1, \tau_2]_{ra}$ is a countable set which has a measure zero. The desired result follows from $\mathcal{M}_l\big([\tau_1, \tau_2]\big) = \tau_1 - \tau_1$.

Q4(4) Prove Eqs. (4.18) to (4.23).

SQ4(4) Use the additive property of measures and the defining properties of a Lebesgue-Stieltjes measure $\mathcal{M}_{ls,g}$ in Eqs. (4.16) and (4.17), i.e., the Lebesgue-Stieltjes measure of a half-open interval $(\tau_1, \tau_2]$ is equal to $g(\tau_2) - g(\tau_1)$ and of a singleton set $\{\tau\}$ is equal to is $g(\tau + 0) - g(\tau - 0)$.

(1) The proof of Eq. (4.18) is the same as the proof of Eq. (3.25) in SQ3(1).

(2) Eq. (4.19) is part of the defining properties of the measure given in Eq. (4.17), bearing in mind that $g(\tau + 0) = g(\tau)$.

(3) Eq. (4.20) is part of the defining properties of the measure given in Eq. (4.16).

(4) To prove of Eq. (4.21) we first observe that

$$\begin{aligned}\mathcal{M}_{ls,g}\big((\tau_1, \tau_2]\big) &= \mathcal{M}_{ls,g}\big((\tau_1, \tau_2) \cup \{\tau_2\}\big)\\ &= \mathcal{M}_{ls,g}\big((\tau_1, \tau_2)\big) + \mathcal{M}_{ls,g}\big(\{\tau_2\}\big)\\ &= \mathcal{M}_{ls,g}\big((\tau_1, \tau_2)\big) + g(\tau_2) - g(\tau_2 - 0).\end{aligned}$$

It follows that

$$\begin{aligned}\mathcal{M}_{ls,g}\big((\tau_1, \tau_2)\big) &= \mathcal{M}_{ls,g}\big((\tau_1, \tau_2]\big) - \big(g(\tau_2) - g(\tau_2 - 0)\big)\\ &= g(\tau_2 - 0) - g(\tau_1).\end{aligned}$$

(5) We can prove of Eq. (4.22) as follows:

$$\begin{aligned}\mathcal{M}_{ls,g}\big([\tau_1, \tau_2]\big) &= \mathcal{M}_{ls,g}\big(\{\tau_1\} \cup (\tau_1, \tau_2]\big)\\ &= \mathcal{M}_{ls,g}\big(\{\tau_1\}\big) + \mathcal{M}_{ls,g}\big((\tau_1, \tau_2]\big)\\ &= g(\tau_1) - g(\tau_1 - 0) + g(\tau_2) - g(\tau_1)\\ &= g(\tau_2) - g(\tau_1 - 0).\end{aligned}$$

(6) We can prove of Eq. (4.23) we first observe that:

$$\begin{aligned}\mathcal{M}_{ls,g}\big([\tau_1, \tau_2]\big) &= \mathcal{M}_{ls,g}\big([\tau_1, \tau_2) \cup \{\tau_2\}\big)\\ &= \mathcal{M}_{ls,g}\big([\tau_1, \tau_2)\big) + \mathcal{M}_{ls,g}\big(\{\tau_2\}\big)\\ &= \mathcal{M}_{ls,g}\big([\tau_1, \tau_2)\big) + \big(g(\tau_2) - g(\tau_2 - 0)\big).\end{aligned}$$

It follows that

$$\begin{aligned}\mathcal{M}_{ls,g}\big([\tau_1, \tau_2)\big) &= \mathcal{M}_{ls,g}\big([\tau_1, \tau_2]\big) - \big(g(\tau_2) - g(\tau_2 - 0)\big)\\ &= g(\tau_2 - 0) - g(\tau_1 - 0).\end{aligned}$$

Q4(5) Explain the main differences between Riemann and Lebesgue integrals.

SQ4(5) The main difference lies in the way the area under the curve is partitioned into a sum of rectangular areas. For Riemann integrals we first partition the domain of the function (integrand) into subintervals Λ_ℓ and then partition the area under the curve into a sum of rectangles with Λ_ℓ as the base. For Lebesgue integrals we partition the range of the function (integrand) into subintervals Λ'_ℓ and take the inverse images $\Lambda_\ell = f^{-1}(\Lambda'_\ell)$ under the function. For Borel functions these inverse images are necessarily Borel sets which may not be intervals. We then take the Lebesgue measure of each of these inverse images $\mathcal{M}_\ell(\Lambda'_\ell)$ as the base to form an area which is not necessarily of the form of a standard rectangle. The area under the curve is approximated by the sum of these areas. The two integrals agree for some functions but not all functions. Lebesgue integrals are applicable to a wider class of functions.

Q4(6) Explain why Lebesgue integrals are only defined for Borel functions.

SQ4(6) By dividing the range of the function (integrand) into subintervals we need to take inverse images of the subintervals under the function (integrand) and we have to give each of these inverse images a length. If these inverses are Borel sets then we can take the Lebesgue measure of these inverse images to give each of these inverse images a length. The integrand being a Borel function ensures that all inverses are Borel sets. Otherwise we will be unable to give each of these inverse images a length, i.e., we are unable to obtain a area associated with each of these inverse images and we are unable to obtain an approximation to the area under the curve.

Q4(7) Explain the main differences between Riemann–Stieltjes and Lebesgue–Stieltjes integrals.

SQ4(7) The Riemann-Stieltjes integrals are extensions of Riemann integrals by integrating with respect to a function rather than an independent variable. The Lebesgue-Stieltjes integrals are extensions of Lebesgue integrals by using a Lebesgue-Stieltjes measure to construct the integrals.

Q4(8) Explain the meaning of Eq. (4.68) which expresses the Dirac delta function $\delta(\tau)$ formally as the derivative of the unit step function $g_{\text{us}}(\tau)$.

SQ4(8) The meaning of Eq. (4.68) which expresses the Dirac delta function $\delta(\tau)$ formally as the derivative of the unit step function $g_{us}(\tau)$, i.e., $\delta(\tau) = dg_{us}(\tau)/d\tau$, lies in Eq. (4.69) on the Riemann-Stieltjes integral with respect to the unit step function. Indeed the

Dirac delta function is meaningful only within the context of such an integral.

Q4(9) Show that the coefficients c_1, c_2, c_3 in Eq. (4.79) must satisfy the condition $c_1 + c_2 + c_3 = 1$.

SQ4(9) Using the limiting value of distribution functions $\mathcal{F}(\tau)$ at $\tau = \infty$ in Definition 3.6(2), i.e., property MP3.6(2), we get

$$\mathcal{F}(\infty) = \mathcal{F}_d(\infty) = \mathcal{F}_{ac}(\infty) = \mathcal{F}_{sc}(\infty) = 1.$$

For a general probability distribution function $\mathcal{F}(\tau)$ in Eq. (4.79) we have

$$\mathcal{F}(\infty) = c_1\mathcal{F}_d(\infty) + c_2\mathcal{F}_{ac}(\infty) + c_3\mathcal{F}_{sc}(\infty) = c_1 + c_2 + c_3.$$

The fact that we must also have $\mathcal{F}(\infty) = 1$ implies $c_1 + c_2 + c_3 = 1$.

Q4(10) Show that expectation value is additive in the sense that for the distribution function in Eq. (4.79) we have

$$\begin{aligned}&\mathcal{E}(c_1\mathcal{F}_{\text{d}} + c_2\mathcal{F}_{\text{ac}} + c_3\mathcal{F}_{\text{sc}})\\ &= c_1\mathcal{E}(\mathcal{F}_{\text{d}}) + c_2\mathcal{E}(\mathcal{F}_{\text{ac}}) + c_3\mathcal{E}(\mathcal{F}_{\text{sc}}).\end{aligned}$$

SQ4(10)[1] The expectation value $\mathcal{E}(\mathcal{F})$ for the distribution function $\mathcal{F}$ in Eq. (4.79) is given by

$$\begin{aligned}\mathcal{E}(\mathcal{F}) &= \int_{-\infty}^{\infty} \tau d\mathcal{F}(\tau)\\ &= \int_{-\infty}^{\infty} \tau d\Big(c_1\mathcal{F}_d(\tau) + c_2\mathcal{F}_{ac}(\tau) + c_3\mathcal{F}_{sc}(\tau)\Big)\\ &= c_1\int_{-\infty}^{\infty} \tau d\mathcal{F}_d(\tau) + c_2\int_{-\infty}^{\infty} \tau d\mathcal{F}_{ac}(\tau) + c_3\int_{-\infty}^{\infty} \tau d\mathcal{F}_{sc}(\tau)\\ &= c_1\mathcal{E}(\mathcal{F}_d) + c_2\mathcal{E}(\mathcal{F}_{ac}) + c_3\mathcal{E}(\mathcal{F}_{sc}).\end{aligned}$$

Q4(11) Show that the Gaussian probability distribution function given by the density function in Eq. (4.80) by

$$\mathcal{F}_{\text{G}}(\tau) = \int_{-\infty}^{\tau} w_{\text{G}}(\tau)\, d\tau$$

[1] Papoulis p. 143.

satisfies the defining properties of probability distribution functions. Verify that the expectation value and the uncertainty are given respectively by

$$\mathcal{E}(\mathcal{F}_G) = b \quad \text{and} \quad \Delta(\mathcal{F}_G) = a.$$

SQ4(11) The defining properties of probability distribution functions are given by Definition 3.6(2). The function $\mathcal{F}_G(\tau)$, being an integral of the Gaussian probability density function, is clearly nondecreasing, continuous from the right and having the limiting value $\mathcal{F}_G(-\infty) = 0$. For the limiting value at ∞ we have

$$\frac{1}{\sqrt{2\pi a^2}} \int_{-\infty}^{\infty} e^{-(\tau-b)^2/2a^2}\, d\tau = 1,$$

due to the well-known integral

$$\int_{-\infty}^{\infty} e^{-\lambda(x-c)^2}\, dx = \sqrt{\frac{\pi}{\lambda}}.$$

The expectation value is calculated by replacing τ by $x = \tau - b$, i.e.,

$$\begin{aligned}
\mathcal{E}(\mathcal{F}_G) &= \frac{1}{\sqrt{2\pi a^2}} \int_{-\infty}^{\infty} \tau\, e^{-(\tau-b)^2/2a^2}\, d\tau \\
&= \frac{1}{\sqrt{2\pi a^2}} \int_{-\infty}^{\infty} (x+b)\, e^{-x^2/2a^2}\, dx \\
&= \frac{1}{\sqrt{2\pi a^2}} \int_{-\infty}^{\infty} x\, e^{-x^2/2a^2}\, dx + \frac{1}{\sqrt{2\pi a^2}} \int_{-\infty}^{\infty} b\, e^{-x^2/2a^2}\, dx \\
&= b.
\end{aligned}$$

The first integral vanishes since the integrand is an odd function of x.

The variance is calculated using Eq. (4.92), i.e.,

$$\begin{aligned}
\text{Var}(\mathcal{F}_G) &= \frac{1}{\sqrt{2\pi a^2}} \int_{-\infty}^{\infty} \tau^2\, e^{(\tau-b)^2/2a^2}\, d\tau - \mathcal{E}(\mathcal{F}_G)^2 \\
&= \frac{1}{\sqrt{2\pi a^2}} \int_{-\infty}^{\infty} \tau^2\, e^{-(\tau-b)^2/2a^2}\, d\tau - b^2 \\
&= \frac{1}{\sqrt{2\pi a^2}} \int_{-\infty}^{\infty} (x+b)^2\, e^{-x^2/2a^2}\, dx - b^2 \\
&= \frac{1}{\sqrt{2\pi a^2}} \int_{-\infty}^{\infty} (x^2 + 2xb + b^2)\, e^{-x^2/2a^2}\, dx - b^2 \qquad (*)
\end{aligned}$$

$$= \frac{1}{\sqrt{2\pi a^2}} \int_{-\infty}^{\infty} x^2 \, e^{-x^2/2a^2} \, dx$$
$$= a^2,$$

due to the integral

$$\int_{-\infty}^{\infty} x^2 \, e^{-\lambda x^2} \, dx = \frac{1}{2\lambda} \sqrt{\frac{\pi}{\lambda}}.$$

The uncertainty is then $\Delta(\mathcal{F}_G) = \sqrt{\mathrm{Var}(\mathcal{F}_G)} = a$. Note that there are three integrals in the expression $(*)$ above. The middle integral is zero on account of an odd integrand. The last integral is equal to b^2 since the integral of $\mathcal{F}_G(\tau)$ is equal to 1.

Q4(12) Find the probability distribution function $\mathcal{F}_U(\tau)$ given by the density function $w_U(\tau)$ in Eq. (4.81).

SQ4(12) The uniform probability distribution function is given by

$$\mathcal{F}_U(\tau) = \int_{-\infty}^{\tau} w_U(\tau')\, d\tau' = \begin{cases} 0 & \tau \le a \\ (\tau - a)/(b - a) & x \in (a, b) \\ 1 & x \ge b \end{cases}.$$

Q4(13) Discuss the mathematical and physical differences between discrete and continuous probability distributions.

SQ4(13) Mathematically, a discrete probability distribution function is piecewise-constant with many discontinuities while a continuous probability distribution function is absolutely continuous (we ignore singularly continuous distributions). Physically, the probability of obtaining a single precise value in a statistical experiment with a continuous probability distribution is zero. Any measurement result has to be specified in terms of a Borel set or interval of non-zero measure. This is not the case for a discrete distribution.

Chapter 5

Quantum Mechanical Systems

Q5(1) Explain why the values of a classical observable are deemed to be objective while the values of a quantum observable are generally regarded as non-objective.

SQ5(1) As stated in §2.1.3 and §2.1.4 a kinematic observable of a classical mechanical system is defined to be a numerical function of the state. It follows that a kinematic observable of a classical mechanical system possesses a value in an arbitrary state at any instance of time. If we wish to know what this value is we can perform a measurement which will *reveal* the value of the observable at that time. A measurement can be executed within a short period of time without disturbing the system significantly so that the value revealed by a measurement is the same as the value the observable has before, during and after the measurement with arbitrary accuracy. This is why we say that classical observables have objective values independent of measurement. On account of QMP5.3(3) quantum observables do not have the above properties. Generally the values of a quantum observable are not objective in the following sense. An observable of a quantum system do not possesses a value in an arbitrary state. A measurement would yield a value but this value cannot be assumed to be the value

Quantum Mechanics: Problems and Solutions
K. Kong Wan

ISBN 978-981-4800-72-3 (Paperback), 978-0-429-29647-5 (eBook)
www.jennystanford.com

possessed by the system before the measurement. A repetition of the experiment to measure the observable in the same state may well yield a different value. This is why we say the values of quantum observables are generally non-objective.

Q5(2) Give a brief account of the relationship between states and possessed values of discrete observables of a quantum system.

SQ5(2) The relationship between states and possessed values of observables is contained in QMP5.3(2) which says that there are states in which a given discrete observable has a definite value which can be revealed by measurement. Such values are by definition the value possessed by the observable in those states. A quantum observable does not possess a definite value in an arbitrary state.

Q5(3) Explain why quantum observables cannot be related to the state in the same way kinematic observables of a classical system are related to the state.

SQ5(3) As stated in §2.1.4 a kinematic observable of a classical mechanical system is defined to be a numerical function of the state. As a result a kinematic observable of a classical mechanical system possesses a value in an arbitrary state at any instance of time. A given state automatically determines simultaneously the values of all these observables. QMP5.3(1) tells us that the values of quantum observables are not simultaneously determinable and QMP5.3(3) informs us that a state cannot determine the values of an arbitrary observable. It follows that quantum observables cannot be related to the state in the same way classical kinematic observables are related to the state, i.e., quantum observables cannot be defined as numerical functions of the state.

Q5(4) Discuss the effect of QMP5.3(1) on the specification of states.

SQ5(4) QMP5.3(1) implies that we cannot determine the state of a quantum system by the simultaneous values of all its observables (obtained by simultaneous measurements of all the observables). Instead we have to determine a state by the simultaneous values of an appropriate set of compatible observables.

Q5(5) Explain what is meant by the behaviour of quantum systems being intrinsically probabilistic.

SQ5(5) For classical systems the probabilistic behaviour is due to a lack of a maximum knowledge of the system. Quantum systems behave probabilistically despite a maximum knowledge of the system, as stated in QMP5.3(3). In particular a state cannot determine the values of all observables. Instead a state can only determine of the probability distributions of the values of an arbitrary observable. We call such behaviour intrinsically probabilistic.

Chapter 6

Three-Dimensional Real Vectors

Q6(1) Prove Eqs. (6.1) to (6.3).

SQ6(1)

(1) To prove Eq. (6.1) we start with the premise $\vec{v}_1 + \vec{u} = \vec{v}_2 + \vec{u}$. Adding $\vec{u}^{-1}$ on both side this equation we get

$$\vec{v}_1 + \vec{u} + \vec{u}^{-1} = \vec{v}_2 + \vec{u} + \vec{u}^{-1}$$
$$\Rightarrow \vec{v}_1 + \vec{0} = \vec{v}_2 + \vec{0} \Rightarrow \vec{v}_1 = \vec{v}_2.$$

(2) To prove the first equation in Eq. (6.2) we start with

$$a\vec{u} = (a+0)\vec{u} = a\vec{u} + 0\vec{u}.$$

Adding $(a\vec{u})^{-1}$ on both side of the equation we get $\vec{0} = 0\vec{u}$. To prove the second equation in Eq. (6.2) we start with

$$a\vec{0} + a\vec{u} = a(\vec{0} + \vec{u}) = a\vec{u} = \vec{0} + a\vec{u}.$$

It then follows from Eq. (6.1) that $a\vec{0} = \vec{0}$.

(3) Eq. (6.3) is proved as follows:

$$\vec{u} + (-1)\vec{u} = \big(1 + (-1)\big)\vec{u} = 0\vec{u} = \vec{0} \Rightarrow (-1)\vec{u} = \vec{u}^{-1}.$$

Quantum Mechanics: Problems and Solutions
K. Kong Wan

ISBN 978-981-4800-72-3 (Paperback), 978-0-429-29647-5 (eBook)
www.jennystanford.com

Q6(2) Prove Theorem 6.2.1(1).

SQ6(2) Let c'_ℓ be a different set of coefficients such that

$$\vec{v} = \sum_{\ell=1}^{n} c'_\ell \vec{u}_\ell.$$

Subtract Eq. (6.9) from the above equation we get

$$\vec{0} = \sum_{\ell=1}^{n} (c_\ell - c'_\ell)\vec{u}_\ell.$$

On account of the linear independence of $\vec{u}_\ell$ the above equation implies that $c_\ell - c'_\ell = 0 \ \forall \ell$. In other words we have $c'_\ell = c_\ell$, i.e., there is just one set of coefficients c_ℓ for any given $\vec{v}$.

Q6(3) Verify that the expression for $\langle \vec{u} \mid \vec{v} \rangle$ in Eq. (6.13) satisfies properties SP6.3.1(1), SP6.3.1(2) and SP6.3.1(3) of scalar product.

SQ6(3)

(1) To prove SP6.3.1(1) we note that changing the order of $\vec{u}$ and $\vec{v}$ would change the angle between the two vectors from θ to $-\theta$. But this does not change the value of the scalar product since $\cos\theta = \cos(-\theta)$, i.e., property SP6.3.1(1) is satisfied.

(2) The distributive property SP6.3.1(2) can be verified diagrammatically. Draw a diagram on the x-y plane with the vector $\vec{u}$ lying along the x-axis and then draw $a_1\vec{v}_1$, $a_2\vec{v}_2$ and their sum $a_1\vec{v}_1 + a_2\vec{v}_2$. Then draw a diagram of projections of $a_1\vec{v}_1$, $a_2\vec{v}_2$ and $a_1\vec{v}_1 + a_2\vec{v}_2$, onto $\vec{u}$, i.e., onto the x-axis. It will then become obvious that the component of $a_1\vec{v}_1 + a_2\vec{v}_2$ onto $\vec{u}$ is equal to the sum of the components of $a_1\vec{v}_1$ and $a_2\vec{v}_2$ onto $\vec{u}$. In other words the distributive property SP6.3.1(2) is satisfied.

(3) Property SP6.3.1(3) is satisfied since $\langle \vec{u} \mid \vec{u} \rangle = u^2 \geq 0$, and $\langle \vec{u} \mid \vec{u} \rangle = 0 \Rightarrow u = 0 \Rightarrow \vec{u} = \vec{0}$. Formally this is because $\vec{u}$ is equal to its norm $||\vec{u}\,||$ times its unit directional vector $\vec{u}^{(u)}$, i.e., $\vec{u} = ||\vec{u}\,||\,\vec{u}^{(u)}$. Then $||\vec{u}|| = 0$ implies $\vec{u} = 0\,\vec{u}^{(u)} = \vec{0}$.

Q6(4) Show that two orthogonal vectors are linearly independent.

SQ6(4) Let $\vec{v}_1$ and $\vec{v}_2$ be two (non-zero) orthogonal vectors and a_1 and a_2 be two real numbers such that $a_1\vec{v}_1 + a_2\vec{v}_2 = \vec{0}$. Then:

$$\langle \vec{v}_1 \mid a_1\vec{v}_1 + a_2\vec{v}_2 \rangle = 0$$

$$\Rightarrow \quad \langle \vec{v}_1 \mid a_1\vec{v}_1 + a_2\vec{v}_2 \rangle = a_1\langle \vec{v}_1 \mid \vec{v}_1 \rangle + a_2\langle \vec{v}_1 \mid \vec{v}_2 \rangle$$

$$= a_1\langle \vec{v}_1 \mid \vec{v}_1 \rangle = 0$$

$$\Rightarrow \quad a_1 = 0 \quad \Rightarrow \quad a_2 = 0.$$

The two vectors are therefore linearly independent.

One can further check this by assuming linear dependence, i.e., $\vec{v}_1 = a\,\vec{v}_2$. We will then have

$$\langle \vec{v}_1 \mid \vec{v}_1 \rangle = a\,\langle \vec{v}_1 \mid \vec{v}_2 \rangle = 0,$$

which is a contradiction since $\langle \vec{v}_1 \mid \vec{v}_1 \rangle = ||\,\vec{v}_1||^2 \neq 0$.

Q6(5) Prove Eq. (6.19).

SQ6(5) Take the scalar product of $\vec{e}_\ell$ and $\vec{v}$ we get

$$\langle \vec{e}_\ell \mid \vec{v} \rangle = \langle \vec{e}_\ell \mid \sum_{\ell'=1}^{3} c_{\ell'}\vec{e}_{\ell'} \rangle = \sum_{\ell'=1}^{3} c_{\ell'}\langle \vec{e}_\ell \mid \vec{e}_{\ell'} \rangle$$

$$= \sum_{\ell'=1}^{3} c_{\ell'}\delta_{\ell\ell'} = c_\ell.$$

Note that we have rewritten Eq. (6.19) as $\sum_{\ell'=1}^{3} c_{\ell'}\vec{e}_{\ell'}$ rather than the original expression $\sum_{\ell=1}^{3} c_\ell\vec{e}_\ell$ which will cause confusion when taking the scalar product with $\vec{e}_\ell$. If one wishes to use the original expression one should take the scalar product with $\vec{e}_{\ell'}$ which will show that $c_{\ell'} = \langle \vec{e}_{\ell'} \mid \vec{v} \rangle$.

Q6(6) Prove the Pythagoras theorem in the forms of Eqs. (6.22) and (6.27).

SQ6(6) For the Pythagoras theorem in Eq. (6.22) we have

$$||\vec{v}\,||^2 = \langle \vec{v} \mid \vec{v} \rangle = v_x v_x\,\langle \vec{i} \mid \vec{i} \rangle + v_x v_y\,\langle \vec{i} \mid \vec{j} \rangle + \cdots.$$

The orthonormality of the basis vectors $\vec{i}$, $\vec{j}$, $\vec{k}$ immediately leads to the desired result.

For the Pythagoras theorem in Eq. (6.27) we have

$$\langle \vec{u} \mid \vec{v} \rangle = \langle \sum_{\ell=1}^{3} u_\ell\vec{e}_\ell \mid \sum_{\ell'=1}^{3} v_{\ell'}\vec{e}_{\ell'} \rangle = \sum_{\ell,\ell'=1}^{3} u_\ell v_{\ell'}\langle \vec{e}_\ell \mid \vec{e}_{\ell'} \rangle.$$

The orthonormality of the basis vectors, i.e., $\langle \vec{e}_\ell \mid \vec{e}_{\ell'} \rangle = \delta_{\ell\ell'}$, immediately leads to the desired result.

Q6(7) Prove Eqs. (6.30) and (6.31).

SQ6(7) For Eq. (6.30) we first express $\vec{v}$ in terms of $\vec{e}_\ell$ as in Eq. (6.19), i.e., $\vec{v} = \sum_{\ell=1}^{3} v_\ell \vec{e}_\ell$ with $v_\ell = \langle \vec{e}_\ell \mid \vec{v} \rangle$. Then:

$$v_\ell = \langle \vec{e}_\ell \mid \vec{v} \rangle = 0 \ \forall \ell \ \Rightarrow \ \vec{v} = \vec{0}.$$

For Eq. (6.31) we observe that

$$\langle \vec{e}_\ell \mid \vec{u} \rangle = \langle \vec{e}_\ell \mid \vec{v} \rangle \ \forall \ell \ \Rightarrow \ \langle \vec{e}_\ell \mid \vec{u} - \vec{v} \rangle = 0 \ \forall \ell.$$

Then Eq. (6.30) proved above implies $\vec{u} - \vec{v} = \vec{0}$, i.e., $\vec{u} = \vec{v}$.

Q6(8) Verify the Gram-Schmidt orthogonalisation procedure given in §6.3.5.

SQ6(8) For the Gram-Schmidt orthogonalisation we have:

$$\langle \vec{u}'_1 \mid \vec{u}'_2 \rangle = \langle \vec{v}_1 \mid \vec{v}_2 \rangle - \frac{\langle \vec{v}_1 \mid \vec{v}_2 \rangle}{\langle \vec{v}_1 \mid \vec{v}_1 \rangle} \langle \vec{v}_1 \mid \vec{v}_1 \rangle = 0,$$

$$\langle \vec{u}'_1 \mid \vec{u}'_3 \rangle = \langle \vec{v}_1 \mid \vec{v}_3 \rangle - \frac{\langle \vec{v}_1 \mid \vec{v}_3 \rangle}{\langle \vec{v}_1 \mid \vec{v}_1 \rangle} \langle \vec{v}_1 \mid \vec{v}_1 \rangle - \frac{\langle \vec{u}'_2 \mid \vec{v}_3 \rangle}{\langle \vec{u}'_2 \mid \vec{u}'_2 \rangle} \langle \vec{u}'_1 \mid \vec{u}'_2 \rangle = 0,$$

$$\langle \vec{u}'_2 \mid \vec{u}'_3 \rangle = \langle \vec{u}'_2 \mid \vec{v}_3 \rangle - \frac{\langle \vec{u}'_1 \mid \vec{v}_3 \rangle}{\langle \vec{u}'_1 \mid \vec{u}'_1 \rangle} \langle \vec{u}'_2 \mid \vec{u}'_1 \rangle - \frac{\langle \vec{u}'_2 \mid \vec{v}_3 \rangle}{\langle \vec{u}'_2 \mid \vec{u}'_2 \rangle} \langle \vec{u}'_2 \mid \vec{u}'_2 \rangle = 0.$$

We have used the results $\langle \vec{u}'_1 \mid \vec{u}'_2 \rangle = 0$ and $\langle \vec{u}'_2 \mid \vec{u}'_1 \rangle = 0$.

Q6(9) Prove triangle inequalities (6.38) and (6.39).

SQ6(9) For the triangle inequality in expression (6.38) we have, using the Schwarz inequality, i.e., $|\langle \vec{u} \mid \vec{v} \rangle| \leq ||\vec{u}|| \, ||\vec{v}||$,

$$||\vec{u} + \vec{v}||^2 = \langle \vec{u} + \vec{v} \mid \vec{u} + \vec{v} \rangle = \langle \vec{u} \mid \vec{u} \rangle + \langle \vec{v} \mid \vec{v} \rangle + \langle \vec{u} \mid \vec{v} \rangle + \langle \vec{v} \mid \vec{u} \rangle$$

$$\leq ||\vec{u}||^2 + ||\vec{v}||^2 + 2\,||\vec{u}|| \, ||\vec{v}|| = \big(||\vec{u}|| + ||\vec{u}||\big)^2$$

$$\Rightarrow ||\vec{u} + \vec{v}|| \leq ||\vec{u}|| + ||\vec{u}||.$$

For the triangle inequality in expression (6.39) we have, using the above inequality,

$$||\vec{u}|| = ||\vec{u} - \vec{v} + \vec{v}|| \leq ||\vec{u} - \vec{v}|| + ||\vec{v}||$$

$$\Rightarrow ||\vec{u}|| - ||\vec{v}|| \leq ||\vec{u} - \vec{v}||.$$

Similarly we have

$$||\vec{v}\,|| = ||\vec{v} - \vec{u} + \vec{u}\,|| \leq ||\vec{v} - \vec{u}\,|| + ||\vec{u}\,||$$
$$\Rightarrow ||\,\vec{v}\,|| - ||\,\vec{u}\,|| \leq ||\vec{v} - \vec{u}\,|| = ||\vec{u} - \vec{v}\,||.$$

It follows that

$$\big|||\,\vec{u}\,|| - ||\,\vec{v}\,||\big| \leq ||\vec{u} - \vec{v}\,||.$$

Q6(10) Let $\{\vec{e}_\ell\}$ be an orthonormal basis. Show that any vector $\vec{v}$ is expressible as a sum of the projections $\vec{v}_{\vec{e}_\ell}$ of $\vec{v}$ onto the basis vectors $\vec{e}_\ell$, i.e., $\vec{v} = \vec{v}_{\vec{e}_1} + \vec{v}_{\vec{e}_2} + \vec{v}_{\vec{e}_3}$.

SQ6(10) Since $\vec{e}_\ell$ form an orthonormal basis we have, by Eqs. (6.19) and (6.61),

$$\vec{v} = v_1\vec{e}_1 + v_2\vec{e}_2 + v_3\vec{e}_3 = \vec{v}_1 + \vec{v}_2 + \vec{v}_3,$$

where $\vec{v}_1 = v_1\vec{e}_1$, $\vec{v}_2 = v_2\vec{e}_2$ and $\vec{v}_3 = v_1\vec{e}_3$ are the projections of $\vec{v}$ onto the basis vectors $\vec{e}_\ell$ in accordance with Eq. (6.46).

Chapter 7

Matrices and Their Relations with Vectors

Q7(1) Verify Eq. (7.37) on the trace of square matrices.

SQ7(1) Since

$$\left(\boldsymbol{M}\cdot\boldsymbol{N}\right)_{mn}=\sum_{k}\boldsymbol{M}_{mk}\boldsymbol{N}_{kn}$$

$$\Rightarrow \operatorname{tr}\left(\boldsymbol{M}\cdot\boldsymbol{N}\right)=\sum_{n}\left(\sum_{k}\boldsymbol{M}_{nk}\boldsymbol{N}_{kn}\right).$$

The same result is obtained if we interchange $\boldsymbol{M}$ and $\boldsymbol{N}$, i.e., we get $\operatorname{tr}\left(\boldsymbol{M}\cdot\boldsymbol{N}\right)=\operatorname{tr}\left(\boldsymbol{N}\cdot\boldsymbol{M}\right)$.

Q7(2) Show that the inverse of a square invertible matrix is unique.

SQ7(2) Let $\boldsymbol{M}^{-1}$ be an inverse of $\boldsymbol{M}$ and let $\boldsymbol{N}$ be another inverse of $\boldsymbol{M}$, i.e., we have

$$\boldsymbol{M}^{-1}\cdot\boldsymbol{M}=\boldsymbol{M}\cdot\boldsymbol{M}^{-1}=\boldsymbol{I}\ \text{ and }\ \boldsymbol{N}\cdot\boldsymbol{M}=\boldsymbol{M}\cdot\boldsymbol{N}\pm\boldsymbol{I}.$$

Then

$$\boldsymbol{M}^{-1}=\boldsymbol{M}^{-1}\cdot\boldsymbol{I}=\boldsymbol{M}^{-1}\cdot\left(\boldsymbol{M}\cdot\boldsymbol{N}\right)=\left(\boldsymbol{M}^{-1}\cdot\boldsymbol{M}\right)\cdot\boldsymbol{N}=\boldsymbol{N}.$$

In other words there is no other inverse.

Quantum Mechanics: Problems and Solutions
K. Kong Wan

ISBN 978-981-4800-72-3 (Paperback), 978-0-429-29647-5 (eBook)
www.jennystanford.com

Q7(3) Prove Eq. (7.71).

SQ7(3) Suppose a matrix $\boldsymbol{N}$ exists such that $\boldsymbol{N} \cdot \boldsymbol{M} = \boldsymbol{I}$. Let $\boldsymbol{C}$ be a column matrix such that $\boldsymbol{M} \cdot \boldsymbol{C} = \boldsymbol{0}$. Then

$$\boldsymbol{N} \cdot \left(\boldsymbol{M} \cdot \boldsymbol{C}\right) = \boldsymbol{0}.$$

We get, because of $\left(\boldsymbol{N} \cdot \boldsymbol{M}\right) = \boldsymbol{I}$,

$$\boldsymbol{N} \cdot \left(\boldsymbol{M} \cdot \boldsymbol{C}\right) = \left(\boldsymbol{N} \cdot \boldsymbol{M}\right) \cdot \boldsymbol{C} = \boldsymbol{C} = \boldsymbol{0}.$$

In other words we have

$$\boldsymbol{N} \cdot \boldsymbol{M} = \boldsymbol{I} \;\text{ and }\; \boldsymbol{M} \cdot \boldsymbol{C} = \boldsymbol{0} \;\Rightarrow\; \boldsymbol{C} = \boldsymbol{0}.$$

This implies $\boldsymbol{M}$ is invertible by property P7.4(3) in §7.4. Now multiplying $\left(\boldsymbol{N} \cdot \boldsymbol{M}\right) = \boldsymbol{I}$ by the inverse $\boldsymbol{M}^{-1}$ from the left we get $\boldsymbol{N} = \boldsymbol{M}^{-1}$. This result implies $\boldsymbol{M} \cdot \boldsymbol{N} = \boldsymbol{M} \cdot \boldsymbol{M}^{-1} = \boldsymbol{I}$.

Q7(4) What is the inverse of a diagonal matrix with non-zero diagonal elements M_{jj}?

SQ7(4) The inverse is also a diagonal matrix with diagonal element M_{jj}^{-1}.

Q7(5) Show that the orthogonal matrix $\boldsymbol{R}_z(\theta_z)$ in Eq. (7.149) is not selfadjoint.

SQ7(5) The matrix $\boldsymbol{R}_z(\theta_z)$ in Eq. (7.149) has real elements. This means that its transpose $\boldsymbol{R}_z^T(\theta_z)$ is equal to its adjoint $\boldsymbol{R}_z^\dagger(\theta_z)$. But $\boldsymbol{R}_z^T(\theta_z)$ not equal to $\boldsymbol{R}_z(\theta_z)$. It follows that its adjoint $\boldsymbol{R}_z^\dagger(\theta_z)$ is different from $\boldsymbol{R}_z(\theta_z)$, i.e., $\boldsymbol{R}_z(\theta_z)$ is not selfadjoint.

Q7(6) Prove Eq. (7.160).

SQ7(6) By Eq. (7.81) we get

$$\langle \boldsymbol{U}\boldsymbol{C} \mid \boldsymbol{U}\boldsymbol{C} \rangle = \langle \boldsymbol{U}^\dagger \boldsymbol{U}\boldsymbol{C} \mid \boldsymbol{C} \rangle, \quad \langle \boldsymbol{U}\boldsymbol{C}_1 \mid \boldsymbol{U}\boldsymbol{C}_2 \rangle = \langle \boldsymbol{U}^\dagger \boldsymbol{U}\boldsymbol{C}_1 \mid \boldsymbol{C}_2 \rangle.$$

Since $\boldsymbol{U}^\dagger \boldsymbol{U} = \boldsymbol{I}$ by Eq. (7.158) we get

$$\langle \boldsymbol{U}^\dagger \boldsymbol{U}\boldsymbol{C} \mid \boldsymbol{C} \rangle = \langle \boldsymbol{C} \mid \boldsymbol{C} \rangle, \quad \langle \boldsymbol{U}^\dagger \boldsymbol{U}\boldsymbol{C}_1 \mid \boldsymbol{C}_2 \rangle = \langle \boldsymbol{C}_1 \mid \boldsymbol{C}_2 \rangle.$$

Q7(7) Verify Eq. (7.162) on simultaneous unitary transformations.

SQ7(7) Using Eqs. (7.81) and (7.158) we get

$$\begin{aligned}\langle \boldsymbol{C}' \mid \boldsymbol{M}'\boldsymbol{C}' \rangle &= \langle \boldsymbol{UC} \mid (\boldsymbol{UMU}^\dagger)(\boldsymbol{UC}) \rangle \\ &= \langle \boldsymbol{UC} \mid \boldsymbol{UM}(\boldsymbol{U}^\dagger \boldsymbol{U})\boldsymbol{C} \rangle \\ &= \langle \boldsymbol{UC} \mid \boldsymbol{UMC} \rangle = \langle \boldsymbol{U}^\dagger \boldsymbol{UC} \mid \boldsymbol{MC} \rangle \\ &= \langle \boldsymbol{C} \mid \boldsymbol{MC} \rangle.\end{aligned}$$

Q7(8) Pauli matrices are denoted by $\sigma_x,\ \sigma_y,\ \sigma_z$.

(a) Verify the six properties of Pauli matrices shown in Eqs. (7.41) to (7.49).

(b) Show that $\left(a_x\sigma_x + a_y\sigma_y + a_z\sigma_z\right)^2 = \left(a_x^2 + a_y^2 + a_z^2\right)\boldsymbol{I}_{2\times 2}$, where $\boldsymbol{I}_{2\times 2}$ is the 2×2 identity matrix and $a_x,\ a_y,\ a_z$ are real numbers.

(c) What are the determinant and trace of each of the Pauli matrices?

(d) What is the inverse of each of Pauli matrices?

(e) Verify that Pauli matrix σ_x satisfies the selfadjointness condition in Eq. (7.163) for the vectors $\boldsymbol{C}_{\vec{\alpha}_z}$ and $\boldsymbol{C}_{\vec{\beta}_z}$ in Eq. (7.118).

(f) Show that the eigenvectors of each Pauli matrix corresponding to the eigenvalues ± 1 given by Eqs. (7.116), (7.117) and (7.118) are orthonormal.

SQ7(8)(a) Pauli matrices are

$$\sigma_x := \begin{pmatrix} 0 & 1 \\ 1 & 0 \end{pmatrix},\quad \sigma_y := \begin{pmatrix} 0 & -i \\ i & 0 \end{pmatrix},\quad \sigma_z := \begin{pmatrix} 1 & 0 \\ 0 & -1 \end{pmatrix}.$$

We have

$$\sigma_x \cdot \sigma_y = \begin{pmatrix} 0 & 1 \\ 1 & 0 \end{pmatrix} \cdot \begin{pmatrix} 0 & -i \\ i & 0 \end{pmatrix} = \begin{pmatrix} i & 0 \\ 0 & -i \end{pmatrix},$$

$$\sigma_y \cdot \sigma_x = \begin{pmatrix} 0 & -i \\ i & 0 \end{pmatrix} \cdot \begin{pmatrix} 0 & 1 \\ 1 & 0 \end{pmatrix} = \begin{pmatrix} -i & 0 \\ 0 & i \end{pmatrix}.$$

It follows that $\sigma_x \cdot \sigma_y - \sigma_y \cdot \sigma_x$ is equal to

$$\sigma_x \cdot \sigma_y - \sigma_y \cdot \sigma_x = 2\begin{pmatrix} i & 0 \\ 0 & -i \end{pmatrix} = 2i\begin{pmatrix} 1 & 0 \\ 0 & -1 \end{pmatrix} = 2i\sigma_z.$$

$$\sigma_x \cdot \sigma_y + \sigma_y \cdot \sigma_x = 2\begin{pmatrix} 0 & 0 \\ 0 & 0 \end{pmatrix} = \begin{pmatrix} 0 & 0 \\ 0 & 0 \end{pmatrix}.$$

Similar calculations prove the remaining commutation and anticommutation relations. The fact that the square of a Pauli matrices is equal to the 2 × 2 identity matrix is also similarly proved.

SQ7(8)(b) Expand the square we get

$$\left(a_x\sigma_x + a_y\sigma_y + a_z\sigma_z\right)^2 = a_x^2\sigma_x^2 + a_y^2\sigma_z^2 + a_z^2\sigma_z^2 + \text{cross terms}.$$

The cross terms are all zero since Pauli matrices anticommute, e.g., $a_x a_y\left(\sigma_x \cdot \sigma_y + \sigma_y \cdot \sigma_x\right) = \mathbf{0}_{2\times 2}$. Next we can verify the fact that the square of each Pauli matrix is the 2 × 2 identity matrix $\boldsymbol{I}_{2\times 2}$, e.g.,

$$\sigma_x^2 := \begin{pmatrix} 0 & 1 \\ 1 & 0 \end{pmatrix}\begin{pmatrix} 0 & 1 \\ 1 & 0 \end{pmatrix} = \begin{pmatrix} 1 & 0 \\ 0 & 1 \end{pmatrix} = \boldsymbol{I}_{2\times 2}.$$

It follows that

$$\left(a_x\sigma_x + a_y\sigma_y + a_z\sigma_z\right)^2 = \left(a_x^2 + a_y^2 + a_z^2\right)\boldsymbol{I}_{2\times 2}.$$

SQ7(8)(c) The determinant of a Pauli matrix is equal to −1, i.e., we have

$$\begin{vmatrix} 0 & 1 \\ 1 & 0 \end{vmatrix} = -1, \quad \begin{vmatrix} 0 & -i \\ i & 0 \end{vmatrix} = -1, \quad \begin{vmatrix} 1 & 0 \\ 0 & -1 \end{vmatrix} = -1.$$

The trace of every Pauli matrix is clearly 0.

SQ7(8)(d) The fact that $\sigma_x \cdot \sigma_x = \boldsymbol{I}_{2\times 2}$ implies that the inverse of σ_x is equal to σ_x. The same result applies to σ_y and σ_z, i.e., we have $\sigma_x^{-1} = \sigma_x, \quad \sigma_y^{-1} = \sigma_y, \quad \sigma_z^{-1} = \sigma_z$.

SQ7(8)(e) We need to evaluate $\langle \boldsymbol{C}_{\vec{\alpha}_z} \mid \sigma_x \boldsymbol{C}_{\vec{\alpha}_z}\rangle$, $\langle \sigma_x \boldsymbol{C}_{\vec{\alpha}_z} \mid \boldsymbol{C}_{\vec{\alpha}_z}\rangle$, $\langle \boldsymbol{C}_{\vec{\alpha}_z} \mid \sigma_x \boldsymbol{C}_{\vec{\beta}_z}\rangle$, $\langle \sigma_x \boldsymbol{C}_{\vec{\alpha}_z} \mid \boldsymbol{C}_{\vec{\beta}_x}\rangle$ and so on. First we have

$$\sigma_x \boldsymbol{C}_{\vec{\alpha}_z} = \begin{pmatrix} 0 & 1 \\ 1 & 0 \end{pmatrix} \cdot \begin{pmatrix} 1 \\ 0 \end{pmatrix} = \begin{pmatrix} 0 \\ 1 \end{pmatrix},$$

and

$$\sigma_x \boldsymbol{C}_{\vec{\beta}_z} = \begin{pmatrix} 0 & 1 \\ 1 & 0 \end{pmatrix} \cdot \begin{pmatrix} 0 \\ 1 \end{pmatrix} = \begin{pmatrix} 1 \\ 0 \end{pmatrix}.$$

Then we have, using Eq. (7.79),

$$\langle \boldsymbol{C}_{\vec{\alpha}_z} \mid \sigma_x \boldsymbol{C}_{\vec{\alpha}_z} \rangle = (1\ \ 0) \cdot \begin{pmatrix} 0 \\ 1 \end{pmatrix} = 0,$$

$$\langle \sigma_x \boldsymbol{C}_{\vec{\alpha}_z} \mid \boldsymbol{C}_{\vec{\alpha}_z} \rangle = (0\ \ 1) \cdot \begin{pmatrix} 1 \\ 0 \end{pmatrix} = 0,$$

$$\langle \boldsymbol{C}_{\vec{\alpha}_z} \mid \sigma_x \boldsymbol{C}_{\vec{\beta}_z} \rangle = (1\ \ 0) \cdot \begin{pmatrix} 1 \\ 0 \end{pmatrix} = 1,$$

$$\langle \sigma_x \boldsymbol{C}_{\vec{\alpha}_z} \mid \boldsymbol{C}_{\vec{\beta}_z} \rangle = (0\ \ 1) \cdot \begin{pmatrix} 0 \\ 1 \end{pmatrix} = 1,$$

$$\langle \boldsymbol{C}_{\vec{\beta}_z} \mid \sigma_x \boldsymbol{C}_{\vec{\alpha}_z} \rangle = (0\ \ 1) \cdot \begin{pmatrix} 0 \\ 1 \end{pmatrix} = 1,$$

$$\langle \sigma_x \boldsymbol{C}_{\vec{\beta}_z} \mid \boldsymbol{C}_{\vec{\alpha}_z} \rangle = (1\ \ 0) \cdot \begin{pmatrix} 1 \\ 0 \end{pmatrix} = 1,$$

$$\langle \boldsymbol{C}_{\vec{\beta}_z} \mid \sigma_x \boldsymbol{C}_{\vec{\beta}_z} \rangle = (0\ \ 1) \cdot \begin{pmatrix} 1 \\ 0 \end{pmatrix} = 0,$$

$$\langle \sigma_x \boldsymbol{C}_{\vec{\beta}_z} \mid \boldsymbol{C}_{\vec{\beta}_z} \rangle = (1\ \ 0) \cdot \begin{pmatrix} 0 \\ 1 \end{pmatrix} = 0.$$

The required results follow immediately.

SQ7(8)(f) For normalisation we have

$$\langle \boldsymbol{C}_{\vec{\alpha}_x} \mid \boldsymbol{C}_{\vec{\alpha}_x} \rangle = \frac{1}{2}(1\ \ 1) \cdot \begin{pmatrix} 1 \\ 1 \end{pmatrix} = 1.$$

$$\langle \boldsymbol{C}_{\vec{\alpha}_y} \mid \boldsymbol{C}_{\vec{\alpha}_y} \rangle = \frac{1}{2}(1\ \ -i) \cdot \begin{pmatrix} 1 \\ i \end{pmatrix} = 1.$$

$$\langle \boldsymbol{C}_{\vec{\alpha}_z} \mid \boldsymbol{C}_{\vec{\alpha}_z} \rangle = (1\ \ 0) \cdot \begin{pmatrix} 1 \\ 0 \end{pmatrix} = 1.$$

Similar calculations show that

$$\langle \boldsymbol{C}_{\vec{\beta}_x} \mid \boldsymbol{C}_{\vec{\beta}_x} \rangle = \langle \boldsymbol{C}_{\vec{\beta}_y} \mid \boldsymbol{C}_{\vec{\beta}_y} \rangle = \langle \boldsymbol{C}_{\vec{\beta}_z} \mid \boldsymbol{C}_{\vec{\beta}_z} \rangle = 1.$$

For orthogonality we have

$$\langle \boldsymbol{C}_{\vec{\alpha}_x} \mid \boldsymbol{C}_{\vec{\beta}_x} \rangle = \frac{1}{2}(1\ \ 1) \cdot \begin{pmatrix} 1 \\ -1 \end{pmatrix} = 0.$$

$$\langle \boldsymbol{C}_{\vec{\alpha}_y} \mid \boldsymbol{C}_{\vec{\beta}_y} \rangle = \frac{1}{2}(1\ \ -i) \cdot \begin{pmatrix} 1 \\ -i \end{pmatrix} = 0.$$

$$\langle \boldsymbol{C}_{\vec{\alpha}_z} \mid \boldsymbol{C}_{\vec{\beta}_z} \rangle = (1\ \ 0) \cdot \begin{pmatrix} 0 \\ 1 \end{pmatrix} = 0.$$

Q7(9) Show that

(a) If $\boldsymbol{P}$ is an $n \times n$ projection matrix then $\boldsymbol{I}_{n\times n} - \boldsymbol{P}$ is also a projection matrix.

(b) If $\boldsymbol{P}$ and $\boldsymbol{Q}$ are $n \times n$ projection matrices then $\boldsymbol{P} \cdot \boldsymbol{Q}$ is also a projection matrix if $\boldsymbol{P}$ and $\boldsymbol{Q}$ commute.

(c) If $\boldsymbol{P}$ and $\boldsymbol{Q}$ are $n \times n$ projection matrices then $\boldsymbol{P} + \boldsymbol{Q}$ is also a projection matrix if $\boldsymbol{P} \cdot \boldsymbol{Q} = \boldsymbol{Q} \cdot \boldsymbol{P} = \boldsymbol{0}$, where $\boldsymbol{0}$ is the $n \times n$ zero matrix.

SQ7(9)(a) First, the matrix $\boldsymbol{I}_{n\times n} - \boldsymbol{P}$ is selfadjoint since $\boldsymbol{P}$ is selfadjoint. Explicitly we have, by Eq. (7.60), that

$$\left(\boldsymbol{I}_{n\times n} - \boldsymbol{P}\right)^{\dagger} = \boldsymbol{I}_{n\times n}^{\dagger} - \boldsymbol{P}^{\dagger} = \boldsymbol{I}_{n\times n} - \boldsymbol{P}.$$

Secondly, the matrix $\boldsymbol{I}_{n\times n} - \boldsymbol{P}$ is idempotent, i.e.,

$$\begin{aligned}\left(\boldsymbol{I}_{n\times n} - \boldsymbol{P}\right)^2 &= \boldsymbol{I}_{n\times n}^2 + \boldsymbol{P}^2 - \boldsymbol{I}_{n\times n} \cdot \boldsymbol{P} - \boldsymbol{P} \cdot \boldsymbol{I}_{n\times n} \\ &= \boldsymbol{I}_{n\times n} - \boldsymbol{P},\end{aligned}$$

since $\boldsymbol{P}^2 = \boldsymbol{P}$ and $\boldsymbol{P} \cdot \boldsymbol{I}_{n\times n} = \boldsymbol{P} \cdot \boldsymbol{I}_{n\times n} = \boldsymbol{P}$. The desired result follows from Definition 7.7.5(1) on projection matrices.

SQ7(9)(b) First, we have

$$(\boldsymbol{P} \cdot \boldsymbol{Q})^{\dagger} = \boldsymbol{Q}^{\dagger} \cdot \boldsymbol{P}^{\dagger} = \boldsymbol{Q} \cdot \boldsymbol{P}, \quad (\boldsymbol{P} \cdot \boldsymbol{Q})^2 = (\boldsymbol{P} \cdot \boldsymbol{Q}) \cdot (\boldsymbol{P} \cdot \boldsymbol{Q}).$$

If $\boldsymbol{Q}$ and $\boldsymbol{P}$ commute then

$$(\boldsymbol{P} \cdot \boldsymbol{Q})^{\dagger} = \boldsymbol{P} \cdot \boldsymbol{Q} \quad \text{and}$$

$$\begin{aligned}(\boldsymbol{P} \cdot \boldsymbol{Q})^2 &= (\boldsymbol{P} \cdot \boldsymbol{Q}) \cdot (\boldsymbol{Q} \cdot \boldsymbol{P}) = \boldsymbol{P} \cdot \boldsymbol{Q} \cdot \boldsymbol{P} = \boldsymbol{P} \cdot \boldsymbol{P} \cdot \boldsymbol{Q} \\ &= \boldsymbol{P} \cdot \boldsymbol{Q}.\end{aligned}$$

In other words $\boldsymbol{P} \cdot \boldsymbol{Q}$ is selfadjoint and idempotent. It follows that $\boldsymbol{P} \cdot \boldsymbol{Q}$ is a projection matrix.

SQ7(9)(c) First, $\boldsymbol{P}+\boldsymbol{Q}$ is clearly selfadjoint. For idempotence we have

$$(\boldsymbol{P} + \boldsymbol{Q})^2 = \boldsymbol{P}^2 + \boldsymbol{Q}^2 + \boldsymbol{P} \cdot \boldsymbol{Q} + \boldsymbol{Q} \cdot \boldsymbol{P} = \boldsymbol{P} + \boldsymbol{Q},$$

since $\boldsymbol{P} \cdot \boldsymbol{Q} = \boldsymbol{Q} \cdot \boldsymbol{P} = \boldsymbol{0}$. It follows that $\boldsymbol{P} + \boldsymbol{Q}$ is a projection matrix.

Q7(10) Show that the 2×2 matrices in Eqs. (7.178) to (7.183) are projection matrices.[1] Find their eigenvectors.

SQ7(10) The matrix $P_{\vec{\alpha}_x}$ in Eq. (7.178) is selfadjoint since it is equal to its adjoint, i.e., it is equal to the complex conjugate of its transpose. All we now need is to prove idempotence. This is confirmed by explicitly calculations, i.e.,

$$P^2_{\vec{\alpha}_x} = \frac{1}{4}\begin{pmatrix} 1 & 1 \\ 1 & 1 \end{pmatrix} \cdot \begin{pmatrix} 1 & 1 \\ 1 & 1 \end{pmatrix} = \frac{1}{4}\begin{pmatrix} 2 & 2 \\ 2 & 2 \end{pmatrix} = P_{\vec{\alpha}_x}.$$

Hence $P_{\vec{\alpha}_x}$ is a projection matrix. There are two orthonormal eigenvectors $C_{\vec{\alpha}_x}$ and $C_{\vec{\beta}_x}$ corresponding to the eigenvalues 1 and 0 respectively, i.e., we have

$$P_{\vec{\alpha}_x} C_{\vec{\alpha}_x} = (C_{\vec{\alpha}_x} C^\dagger_{\vec{\alpha}_x}) C_{\vec{\alpha}_x} = C_{\vec{\alpha}_x} (C^\dagger_{\vec{\alpha}_x} C_{\vec{\alpha}_x}) = C_{\vec{\alpha}_x},$$

$$P_{\vec{\alpha}_x} C_{\vec{\beta}_x} = (C_{\vec{\alpha}_x} C^\dagger_{\vec{\alpha}_x}) C_{\vec{\beta}_x} = C_{\vec{\alpha}_x} (C^\dagger_{\vec{\alpha}_x} C_{\vec{\beta}_x}) = \mathbf{0}_{2\times 2} = 0\, C_{\vec{\beta}_x}.$$

We can also verify these results by explicit calculations in terms of the matrix $P_{\vec{\alpha}_x}$ and the column vectors $C_{\vec{\alpha}_x}$ and $C_{\vec{\beta}_x}$.

Similar calculations show that all the matrices in Eqs. (7.179) to (7.183) are selfadjoint and idempotent, i.e., they are projection matrices. They all have eigenvalues 0 and 1 in accordance with Theorem 7.7.5(1). Explicitly we have

$$P_{\vec{\beta}_x} C_{\vec{\alpha}_x} = (C_{\vec{\beta}_x} C^\dagger_{\vec{\beta}_x}) C_{\vec{\alpha}_x} = 0\, C_{\vec{\alpha}_x},$$

$$P_{\vec{\beta}_x} C_{\vec{\beta}_x} = (C_{\vec{\beta}_x} C^\dagger_{\vec{\beta}_x}) C_{\vec{\beta}_x} = C_{\vec{\beta}_x}.$$

$$P_{\vec{\alpha}_y} C_{\vec{\alpha}_y} = (C_{\vec{\alpha}_y} C^\dagger_{\vec{\alpha}_y}) C_{\vec{\alpha}_y} = C_{\vec{\alpha}_y},$$

$$P_{\vec{\alpha}_y} C_{\vec{\beta}_y} = (C_{\vec{\alpha}_y} C^\dagger_{\vec{\alpha}_y}) C_{\vec{\beta}_y} = 0\, C_{\vec{\beta}_y}.$$

$$P_{\vec{\beta}_y} C_{\vec{\alpha}_y} = (C_{\vec{\beta}_y} C^\dagger_{\vec{\beta}_y}) C_{\vec{\alpha}_y} = 0\, C_{\vec{\alpha}_y},$$

$$P_{\vec{\beta}_y} C_{\vec{\beta}_y} = (C_{\vec{\beta}_y} C^\dagger_{\vec{\beta}_y}) C_{\vec{\beta}_y} = C_{\vec{\beta}_y}.$$

$$P_{\vec{\alpha}_z} C_{\vec{\alpha}_z} = (C_{\vec{\alpha}_z} C^\dagger_{\vec{\alpha}_z}) C_{\vec{\alpha}_z} = C_{\vec{\alpha}_z},$$

$$P_{\vec{\alpha}_z} C_{\vec{\beta}_z} = (C_{\vec{\alpha}_z} C^\dagger_{\vec{\alpha}_z}) C_{\vec{\beta}_z} = 0\, C_{\vec{\beta}_z}.$$

---

[1] Isham p. 91.

$$\boldsymbol{P}_{\vec{\beta}_z}\boldsymbol{C}_{\vec{\alpha}_z} = (\boldsymbol{C}_{\vec{\beta}_z}\boldsymbol{C}^{\dagger}_{\vec{\beta}_z})\boldsymbol{C}_{\vec{\alpha}_z} = 0\,\boldsymbol{C}_{\vec{\alpha}_z},$$
$$\boldsymbol{P}_{\vec{\beta}_z}\boldsymbol{C}_{\vec{\beta}_x} = (\boldsymbol{C}_{\vec{\beta}_z}\boldsymbol{C}^{\dagger}_{\vec{\beta}_x})\boldsymbol{C}_{\vec{\beta}_z} = \boldsymbol{C}_{\vec{\beta}_z}.$$

We have used the orthonormality property of the eigenvectors $\boldsymbol{C}_{\vec{\alpha}_x}, \boldsymbol{C}_{\vec{\beta}_x}, \boldsymbol{C}_{\vec{\alpha}_y}, \boldsymbol{C}_{\vec{\beta}_y}, \boldsymbol{C}_{\vec{\alpha}_z}, \boldsymbol{C}_{\vec{\beta}_z}$ of Pauli matrices proved earlier in SQ7(8)(f).

Chapter 8

Operations on Vectors in $\vec{\mathbb{E}}^3$

Q8(1) Show that an invertible operator has a unique inverse.

SQ8(1) Let $\widehat{A}^{-1}$ be an inverse of $\widehat{A}$ and let $\widehat{B}$ be another inverse of $\widehat{A}$, i.e., we have $\widehat{A}^{-1}\widehat{A} = \widehat{A}\widehat{A}^{-1} = \widehat{I\!I}, \quad \widehat{B}\widehat{A} = \widehat{A}\widehat{B} = \widehat{I\!I}$. Then

$$\widehat{A}^{-1} = \widehat{A}^{-1}\widehat{I\!I} = \widehat{A}^{-1}\big(\widehat{A}\widehat{B}\big) = \big(\widehat{A}^{-1}\widehat{A}\big)\widehat{B} = \widehat{B}.$$

Q8(2) Show that in $\vec{\mathbb{E}}^3$ the condition $\widehat{B}\widehat{A} = \widehat{I\!I}$ is sufficient to imply the invertibility of $\widehat{A}$ with $\widehat{B}$ as its inverse.

SQ8(2) Let $\vec{u}$ be a vector such that $\widehat{A}\vec{u} = \vec{0}$. Then we have $\widehat{B}\widehat{A}\vec{u} = \vec{0}$. But $\widehat{B}\widehat{A} = \widehat{I\!I}$. It follows that we must have $\vec{u} = \vec{0}$. Hence $\widehat{A}$ is invertible by Theorem 8.2.2(1). Let $\widehat{A}^{-1}$ be its inverse, then multiplying the equation $\widehat{B}\widehat{A} = \widehat{I\!I}$ from the left by $\widehat{A}^{-1}$ we immediately get the result that $\widehat{B} = \widehat{A}^{-1}$.

Q8(3) Let $\boldsymbol{A}$ be a matrix representation of an invertible operator $\widehat{A}$. Show that the matrix representation of the inverse $\widehat{A}^{-1}$ of $\widehat{A}$ is the inverse matrix $\boldsymbol{A}^{-1}$ to the matrix $\boldsymbol{A}$.

SQ8(3) The matrix representations $\boldsymbol{M}_{\widehat{A}}$, $\boldsymbol{M}_{\widehat{A}^{-1}}$ of $\widehat{A}$ and $\widehat{A}^{-1}$ in a given basis $\{\vec{e}_\ell\}$ are given by their matrix elements

$$\big(M_{\widehat{A}}\big)_{k\ell} = \langle \vec{e}_k \mid \widehat{A}\vec{e}_\ell\rangle, \quad \big(M_{\widehat{A}^{-1}}\big)_{\ell m} = \langle \vec{e}_\ell \mid \widehat{A}^{-1}\vec{e}_m\rangle.$$

Quantum Mechanics: Problems and Solutions
K. Kong Wan

ISBN 978-981-4800-72-3 (Paperback), 978-0-429-29647-5 (eBook)
www.jennystanford.com

The product $\boldsymbol{M}_{\widehat{A}} \cdot \boldsymbol{M}_{\widehat{A}^{-1}}$ has elements $\left(\boldsymbol{M}_{\widehat{A}} \cdot \boldsymbol{M}_{\widehat{A}^{-1}}\right)_{km}$ given by

$$\begin{aligned}
&\sum_{\ell} \left(M_{\widehat{A}}\right)_{k\ell} \left(M_{\widehat{A}^{-1}}\right)_{\ell m} \\
&= \sum_{\ell} \langle \vec{e}_k \mid \widehat{A}\vec{e}_\ell \rangle \langle \vec{e}_\ell \mid \widehat{A}^{-1}\vec{e}_m \rangle \\
&= \sum_{\ell} \langle \widehat{A}^\dagger \vec{e}_k \mid \vec{e}_\ell \rangle \langle \vec{e}_\ell \mid \widehat{A}^{-1}\vec{e}_m \rangle = \langle \widehat{A}^\dagger \vec{e}_k \mid \widehat{A}^{-1}\vec{e}_m \rangle \\
&= \langle \vec{e}_k \mid \widehat{A}\widehat{A}^{-1}\vec{e}_m \rangle = \langle \vec{e}_k \mid \widehat{I\!I}\,\vec{e}_m \rangle = \delta_{km}.
\end{aligned}$$

It follows that $\boldsymbol{M}_{\widehat{A}} \cdot \boldsymbol{M}_{\widehat{A}^{-1}} = \boldsymbol{I\!I} \quad \Rightarrow \quad \boldsymbol{M}_{\widehat{A}^{-1}} = \boldsymbol{M}_{\widehat{A}}^{-1}$. We have used the Pythagoras theorem in the form of Eq. (6.29).

Q8(4) Let $\{\vec{e}_\ell, \ell = 1, 2, 3\}$ be a basis for $\vec{\mathbb{E}}^3$ and let $\widehat{A}$ be an invertible operator. Show that $\{\vec{e}\,'_\ell = \widehat{A}\vec{e}_\ell, \ell = 1, 2, 3\}$ is also a basis for $\vec{\mathbb{E}}^3$.[1]

SQ8(4) All we need is to show that $\vec{e}\,'_\ell$ are linearly independent. We can start by considering an arbitrary linear combination $\sum_\ell c_\ell \vec{e}\,'_\ell$, i.e.,

$$\sum_{\ell=1}^{3} c_\ell \vec{e}\,'_\ell = \sum_{\ell=1}^{3} c_\ell \widehat{A}\vec{e}_\ell = \widehat{A} \sum_{\ell=1}^{3} c_\ell \vec{e}_\ell.$$

Then

$$\sum_{\ell=1}^{3} c_\ell \vec{e}\,'_\ell = \vec{0} \;\Rightarrow\; \widehat{A} \sum_{\ell=1}^{3} c_\ell \vec{e}_\ell = \vec{0}.$$

On account of $\widehat{A}$ being invertible we have, by Theorem 8.2.2(1),

$$\widehat{A} \sum_{\ell=1}^{3} c_\ell \vec{e}_\ell = \vec{0} \;\Rightarrow\; \sum_{\ell=1}^{3} c_\ell \vec{e}_\ell = \vec{0}.$$

Since $\vec{e}_\ell$ are linearly independent we must have $c_\ell = 0\ \forall \ell$. In other words we have

$$\sum_{\ell=1}^{3} c_\ell \vec{e}\,'_\ell = \vec{0} \;\Rightarrow\; c_\ell = 0\ \ \forall \ell.$$

This immediately implies the linear independence of $\vec{e}\,'_\ell$. It follows that $\{\vec{e}\,'_\ell, \ell = 1, 2, 3\}$ is a basis for $\vec{\mathbb{E}}^3$.

[1] Halmos p. 63.

Q8(5) Show that an operator $\widehat{A}$ is invertible if and only if every vector $\vec{v} \in \vec{E}^3$ can be expressed as $\vec{v} = \widehat{A}\vec{u}$ for some $\vec{u} \in \vec{E}^3$.[2]

SQ8(5) First, suppose $\widehat{A}$ is invertible. Then, by definition, $\widehat{A}$ generates a one-to-one mapping of $\vec{E}^3$ onto $\vec{E}^3$. This means that any vector $\vec{v} \in \vec{E}^3$ can be regarded as a vector in the range of the operator. In other words there is a vector $\vec{u}$ in the domain of the operator such that $\widehat{A}\vec{u} = \vec{v}$.

Secondly, suppose $\widehat{A}$ is such that every vector $\vec{v}$ in $\vec{E}^3$ can be expressed as $\vec{v} = \widehat{A}\vec{u}$ for some vector $\vec{u}$. Let $\{\vec{e}'_\ell\}$ be a basis in $\vec{E}^3$. Then there exists a set of vectors $\vec{e}_\ell$ such that $\vec{e}'_\ell = \widehat{A}\vec{e}_\ell$. Moreover these vectors $\vec{e}_\ell$ are linearly independent, i.e.,

$$\sum_\ell c_\ell \vec{e}_\ell = \vec{0} \;\Rightarrow\; c_\ell = 0 \;\forall \ell.$$

We can prove this as follows:

$$\sum_\ell c_\ell \vec{e}_\ell = \vec{0} \;\Rightarrow\; \widehat{A}\sum_\ell c_\ell \vec{e}_\ell = \vec{0}$$

$$\Rightarrow \sum_\ell c_\ell \widehat{A}\vec{e}_\ell = \sum_\ell c_\ell \vec{e}'_\ell = \vec{0} \Rightarrow \; c_\ell = 0 \;\forall \ell.$$

It follows $\{\vec{e}_\ell\}$ is also a basis for $\vec{E}^3$ and any vector $\vec{v} \in \vec{E}^3$ can be written as $\vec{v} = \sum_\ell v_\ell \vec{e}_\ell$. Then

$$\widehat{A}\vec{v} = \vec{0} \;\Rightarrow\; \sum_\ell v_\ell \widehat{A}\vec{e}_\ell = \sum_\ell v_\ell \vec{e}'_\ell = \vec{0}.$$

Since $\vec{e}'_\ell$ are linearly independent we must have $v_\ell = 0 \;\forall \ell$, i.e., $\vec{v} = \vec{0}$. We can conclude that $\widehat{A}$ is invertible by Theorem 8.2.2(1).

Q8(6) Show that the matrix representation of the adjoint of $\widehat{A}$ is the adjoint matrix $\boldsymbol{M}^\dagger_{\widehat{A}}$ to the matrix $\boldsymbol{M}_{\widehat{A}}$.

SQ8(6) The matrix element of $\boldsymbol{M}_{\widehat{A}^\dagger}$ are given by

$$\left(M_{\widehat{A}^\dagger}\right)_{lm} = \langle \vec{e}_l \mid \widehat{A}^\dagger \vec{e}_m\rangle = \langle \widehat{A}\vec{e}_l \mid \vec{e}_m\rangle = \langle \vec{e}_m \mid \widehat{A}\vec{e}_l\rangle.$$

Since $\left(M_{\widehat{A}}\right)_{lm} = \langle \vec{e}_l \mid \widehat{A}\vec{e}_m\rangle$ and $\left(M_{\widehat{A}}\right)_{ml} = \langle \vec{e}_m \mid \widehat{A}\vec{e}_l\rangle$ we get

$$\left(M_{\widehat{A}^\dagger}\right)_{lm} = \left(M_{\widehat{A}}\right)_{ml}.$$

[2] Halmos pp. 62–63.

This means that the transpose of $\boldsymbol{M}_{\widehat{A}}$ is equal to $\boldsymbol{M}_{\widehat{A}^\dagger}$. We arrive at the desired result since the transpose of a real square matrix is equal the adjoint of the matrix. All quantities are real in $\vec{I\!E}^3$, e.g., $\langle \widehat{A}\vec{e}_l \mid \vec{e}_m \rangle = \langle \vec{e}_m \mid \widehat{A}\vec{e}_l \rangle$.

Q8(7) Let F be a linear functional on $\vec{I\!E}^3$. Show that $F(\vec{0}) = 0$. Prove a similar result for linear operators.[3]

SQ8(7) For a linear functional F we have $F(a\vec{u}) = aF(\vec{u})$. Now, let $a = 0$ we get $F(0\vec{u}) = 0F(\vec{u}) = 0$. Since $0\vec{u} = \vec{0}$ it follows that $F(\vec{0}) = 0$.

For a linear operator $\widehat{A}$ we also have $\widehat{A}(a\vec{u}) = a\widehat{A}\vec{u}$. Now, let $a = 0$ we get $\widehat{A}(0\vec{u}) = 0\widehat{A}\vec{u} = \vec{0}$. Since $0\vec{u} = \vec{0}$ it follows that $\widehat{A}\vec{0} = \vec{0}$.

Q8(8) Prove Eqs. (8.38) to (8.42) on adjoint operations.

SQ8(8)

(1) For Eq. (8.17) and SP6.3.1(2) on scalar product we get, $\forall\, \vec{u}, \vec{v}$,

$$\begin{aligned}
&\langle \vec{u} \mid (a\widehat{A} + b\widehat{B})\vec{v} \rangle \\
&= \langle \vec{u} \mid a\widehat{A}\vec{v} \rangle + \langle \vec{u} \mid b\widehat{B}\vec{v} \rangle = \langle a\widehat{A}^\dagger\vec{u} \mid \vec{v} \rangle + \langle b\widehat{B}^\dagger\vec{u} \mid \vec{v} \rangle \\
&= \langle (a\widehat{A}^\dagger + b\widehat{B}^\dagger)\vec{u} \mid \vec{v} \rangle \;\Rightarrow\; (a\widehat{A} + b\widehat{B})^\dagger = a\widehat{A}^\dagger + b\widehat{B}^\dagger.
\end{aligned}$$

Equation (8.39) is a special case of Eq. (8.38).

(2) To prove Eq. (8.40) we have, using Eq. (8.34),

$$\langle \vec{u} \mid \widehat{A}^{\dagger\dagger}\vec{v} \rangle = \langle \vec{u} \mid (\widehat{A}^\dagger)^\dagger \vec{v} \rangle = \langle \widehat{A}^\dagger\vec{u} \mid \vec{v} \rangle = \langle \vec{u} \mid \widehat{A}\vec{v} \rangle.$$

Since this is true for all $\vec{u}$ and $\vec{v}$ we can conclude that $\widehat{A}^{\dagger\dagger} = \widehat{A}$.

(3) To prove Eq. (8.41) we have, $\forall\, \vec{u}, \vec{v}$,

$$\langle \vec{u} \mid \widehat{A}\widehat{B}\vec{v} \rangle = \langle \widehat{A}^\dagger\vec{u} \mid \widehat{B}\vec{v} \rangle = \langle \widehat{B}^\dagger\widehat{A}^\dagger\vec{u} \mid \vec{v} \rangle \;\Rightarrow\; (\widehat{A}\widehat{B})^\dagger = \widehat{B}^\dagger\widehat{A}^\dagger.$$

(4) To prove Eq. (8.42) we make used of Eq. (8.41) to get

$$(\widehat{A}\widehat{A}^{-1})^\dagger = (\widehat{A}^{-1})^\dagger\widehat{A}^\dagger.$$

Since $\widehat{A}\widehat{A}^{-1} = \widehat{I\!I}$ and $\widehat{I\!I}^\dagger = \widehat{I\!I}$ we get

$$\widehat{I\!I} = (\widehat{A}^{-1})^\dagger\widehat{A}^\dagger \;\Rightarrow\; (\widehat{A}^{-1})^\dagger = (\widehat{A}^\dagger)^{-1}.$$

[3] Halmos pp. 20, 55.

In (2) and (3) above we have used Eq. (6.31) which tells us that

$$\langle \vec{u} \mid \widehat{A}\vec{v}\,\rangle = \langle \vec{u} \mid \widehat{B}\vec{v}\rangle \;\forall\, \vec{u},\, \vec{v} \in \vec{\mathbb{E}}^3 \quad \Rightarrow \quad \widehat{A} = \widehat{B}.$$

Q8(9) Associated with the projection operation given by Eq. (6.41) we can define an operator $\widehat{P}_{\vec{i}}$ by

$$\widehat{P}_{\vec{i}}\,\vec{v} = v_x\,\vec{i} \quad \forall \vec{v} \in \vec{\mathbb{E}}^3.$$

Show that

(a) The operator $\widehat{P}_{\vec{i}}$ is idempotent.

(b) The operator $\widehat{P}_{\vec{i}}$ is selfadjoint.

(c) The operator possesses only two eigenvalues 0 and 1. Find the corresponding eigenvectors.

(d) The matrix representation $\boldsymbol{M}_{\widehat{P}_{\vec{i}}}$ of $\widehat{P}_{\vec{i}}$ in basis $\{\vec{i}, \vec{j}, \vec{k}\}$ is

$$\boldsymbol{M}_{\widehat{P}_{\vec{i}}} = \begin{pmatrix} 1 & 0 & 0 \\ 0 & 0 & 0 \\ 0 & 0 & 0 \end{pmatrix}.$$

Show also that $\boldsymbol{M}_{\widehat{P}_{\vec{i}}}$ is a projection matrix. Find the eigenvalues and their corresponding eigenvectors of $\boldsymbol{M}_{\widehat{P}_{\vec{i}}}$.

SQ8(9)(a) $\big(\widehat{P}_{\vec{i}}\big)^2\vec{v} = \widehat{P}_{\vec{i}}\widehat{P}_{\vec{i}}\,\vec{v} = v_x\widehat{P}_{\vec{i}}\,\vec{i} = v_x\vec{i} \;\; \forall\vec{v} \;\; \Rightarrow \;\; \big(\widehat{P}_{\vec{i}}\big)^2 = \widehat{P}_{\vec{i}}.$

SQ8(9)(b) The operator $\widehat{P}_{\vec{i}}$ is selfadjoint since

$$\langle \vec{u} \mid \widehat{P}_{\vec{i}}\,\vec{v}\,\rangle = \langle \vec{u} \mid v_x\vec{i}\,\rangle = v_x\langle \vec{u} \mid \vec{i}\,\rangle = v_x u_x.$$

$$\langle \widehat{P}_{\vec{i}}\,\vec{u} \mid \vec{v}\,\rangle = \langle u_x\vec{i} \mid \vec{v}\,\rangle = u_x\langle \vec{i} \mid \vec{v}\,\rangle = u_x v_x.$$

$$\Rightarrow \quad \langle \vec{u} \mid \widehat{P}_{\vec{i}}\,\vec{v}\,\rangle = \langle \widehat{P}_{\vec{i}}\,\vec{u} \mid \vec{v}\,\rangle \quad \forall \vec{u},\, \vec{v}.$$

The concept of selfadjointness is introduced for square matrices by Definition 7.7.4(1). A square matrix is selfadjoint if it satisfies that selfadjointness condition stated after Definition 7.7.4(1). We have applied the condition to show that the projector $\widehat{P}_{\vec{i}}$ is selfadjoint. All these features are introduced for projectors in P9.3.1(6) and for operators in Definition 9.4.1(1).

SQ8(9)(c) Let $\vec{e}$ be an eigenvector of $\widehat{P}_{\vec{i}}$ corresponding to eigenvalue a, i.e., $\widehat{P}_{\vec{i}}\,\vec{e} = a\vec{e}$, where a is real (*see* Eq. (8.49)). We have, on account of $\widehat{P}_{\vec{i}}$ being idempotent,

$$\big(\widehat{P}_{\vec{i}}\big)^2\vec{e} = \widehat{P}_{\vec{i}}\,\vec{e} = a\vec{e} \quad \text{and}$$

$$\left(\widehat{P}_{\vec{i}}\right)^2\vec{e} = \widehat{P}_{\vec{i}}\left(\widehat{P}_{\vec{i}}\vec{e}\right) = a\widehat{P}_{\vec{i}}\vec{e} = a^2\vec{e} \;\Rightarrow\; a^2 = a$$

$$\Rightarrow\; a = 0 \text{ or } 1.$$

Eigenvectors corresponding to the eigenvalue 1 is of the form $u_x\vec{i}$ while eigenvectors corresponding to the eigenvalue 0 is of the form $u_y\vec{j} + u_z\vec{k}$. The eigenvalue 1 is nondegenerate since we do not count two eigenvectors differing only by a multiplicative constant as distinct. The eigenvalue 0 is degenerate with degeneracy 2. Any linear combination of $\vec{j}$ and $\vec{k}$ is also an eigenvector of $\widehat{P}_{\vec{i}}$ corresponding to the eigenvalue 0.

SQ8(9)(d) The given matrix $\boldsymbol{M}_{\widehat{P}_{\vec{i}}}$ is clearly selfadjoint and idempotent. It follows that $\boldsymbol{M}_{\widehat{P}_{\vec{i}}}$ is a projection matrix by Definition 7.7.5(1). By the same argument presented for the preceding question we can see that the matrix has two eigenvalues 1 and 0. The eigenvector corresponding to the eigenvalue 1 is

$$\begin{pmatrix} 1 \\ 0 \\ 0 \end{pmatrix},$$

and the eigenvectors corresponding to the eigenvalue 0 are linear combinations of

$$\begin{pmatrix} 0 \\ 1 \\ 0 \end{pmatrix} \quad \text{and} \quad \begin{pmatrix} 0 \\ 0 \\ 1 \end{pmatrix}.$$

The eigenvalue 1 is nondegenerate while the eigenvalue 0 is degenerate with degeneracy 2.

Chapter 9

Special Operators on $\vec{E}^3$

Q9(1) Show that an operator on $\vec{E}^3$ is orthogonal if and only if it is invertible and its inverse is equal to its adjoint.

SQ9(1) Let $\widehat{R}$ be an orthogonal operator. By Definition 9.2.2(1) we have $||\,\widehat{R}\vec{u}\,|| = ||\,\vec{u}\,||$. It follows that

$$\widehat{R}\vec{u} = \vec{0} \;\Rightarrow\; ||\,\vec{u}\,|| = 0 \;\Rightarrow\; \vec{u} = \vec{0}.$$

Theorem 8.2.2(1) then tells us that $\widehat{R}$ is invertible. Next we have, by Eq. (8.34),

$$\langle \vec{u} \mid \vec{u}\,\rangle = \langle \widehat{R}\vec{u} \mid \widehat{R}\vec{u}\,\rangle = \langle \vec{u} \mid \widehat{R}^\dagger \widehat{R}\vec{u}\,\rangle \;\; \forall \vec{u}.$$

As pointed out in P9.4.2(4) the product operator $\widehat{R}^\dagger \widehat{R}$ is selfadjoint. Hence Corollary 9.4.1(1) applies, i.e., we have

$$\widehat{R}^\dagger \widehat{R} = \widehat{I\!I} \Rightarrow \widehat{R}^\dagger = \widehat{R}^{-1}.$$

To prove the converse we first assume that $\widehat{R}$ is invertible with its inverse $\widehat{R}^{-1}$ equal its adjoint $\widehat{R}^\dagger$. Then we have

$$\langle \widehat{R}\vec{u} \mid \widehat{R}\vec{u}\,\rangle = \langle \vec{u} \mid \widehat{R}^\dagger \widehat{R}\vec{u}\,\rangle = \langle \vec{u} \mid \widehat{R}^{-1}\widehat{R}\vec{u}\,\rangle = \langle \vec{u} \mid \vec{u}\,\rangle \;\forall \vec{u}.$$

The operator is therefore orthogonal.

Q9(2) Show that orthogonal operators on $\vec{E}^3$ have at most two eigenvalues, i.e., ± 1. Find the eigenvalues and eigenvectors of $\widehat{R}_{rx}$ in Eq. (9.4) and $\widehat{R}_{rxyz}$ in Eq. (9.5).

Quantum Mechanics: Problems and Solutions
K. Kong Wan

ISBN 978-981-4800-72-3 (Paperback), 978-0-429-29647-5 (eBook)
www.jennystanford.com

SQ9(2) Let r be an eigenvalue an orthogonal operator $\widehat{R}$ corresponding to normalised eigenvector $\vec{e}$, i.e.,

$$\widehat{R}\vec{e} = r\vec{e} \quad \text{where} \quad \langle \vec{e} \mid \vec{e} \rangle = 1.$$

Then

$$\langle \widehat{R}\vec{e} \mid \widehat{R}\vec{e} \rangle = r^2 \quad \text{and} \quad \langle \widehat{R}\vec{e} \mid \widehat{R}\vec{e} \rangle = \langle \vec{e} \mid \vec{e} \rangle = 1$$

$$\Rightarrow \; r^2 = 1 \;\Rightarrow\; r = \pm 1.$$

The operator $\widehat{R}_{rx}$ in Eq. (9.4) has eigenvalue -1, corresponding to eigenvectors of the form $v_x\vec{i}$ and eigenvalue 1 corresponding to eigenvectors of the form $v_y\vec{j} + v_z\vec{k}$.

The operator $\widehat{R}_{rxyz}$ in Eq. (9.5) has only one eigenvalue, i.e., -1, which is degenerate corresponding to eigenvectors of the form $v_x\vec{i} + v_y\vec{i} + v_z\vec{i}$, i.e., any vector in $\vec{\mathbb{E}}^3$. The operator acts like $-\widehat{\mathbb{I}}$.

Q9(3) Let $\{\vec{e}_\ell\}$ be an orthonormal basis for $\vec{\mathbb{E}}^3$ show that their orthogonal transforms $\{\vec{e}\,'_\ell\}$ generated by an orthogonal operator $\widehat{R}$ is also an orthonormal basis for $\vec{\mathbb{E}}^3$.

SQ9(3) An orthogonal transformation preserves the scalar product. It follows that the transformed vectors $\vec{e}\,'_\ell$ of the basis vectors $\vec{e}_\ell$ would remain orthonormal. Since orthogonal vectors are also linearly independent the transformed vectors $\vec{e}_\ell^{\,'}$ would form a complete orthonormal set, i.e., an orthonormal basis.

Q9(4) Prove Eqs. (9.6) to (9.8) on orthogonal transformations.

SQ9(4) We can prove Eqs. (9.6) to (9.8) using the property $\widehat{R}^\dagger \widehat{R} = \widehat{\mathbb{I}}$ shown in SQ9(1).

For Eq. (9.6) we have

$$\langle \widehat{R}\vec{u} \mid \widehat{R}\vec{v} \rangle = \langle \vec{u} \mid \widehat{R}^\dagger \widehat{R}\vec{v} \rangle = \langle \vec{u} \mid \vec{v} \rangle.$$

For Eq. (9.7) we have

$$\begin{aligned}\langle \vec{u}\,' \mid \widehat{A}\,'\vec{v}\,' \rangle &= \langle \widehat{R}\vec{u} \mid \widehat{R}\widehat{A}\widehat{R}^\dagger \widehat{R}\vec{v} \rangle = \langle \widehat{R}\vec{u} \mid \widehat{R}\widehat{A}\vec{v} \rangle \\ &= \langle \widehat{R}^\dagger \widehat{R}\,\vec{u} \mid \widehat{A}\vec{v} \rangle = \langle \vec{u} \mid \widehat{A}\vec{v} \rangle.\end{aligned}$$

Eq. (9.8) follows from Eq. (9.7).

Q9(5) Prove Theorem 9.3.2(1).

SQ9(5) By Definition 9.3.2(1) we have, $\forall \vec{v} \in I\!E^3$,

$$(\widehat{P}_{\vec{S}} + \widehat{P}_{\vec{S}^\perp})\vec{v} = \widehat{P}_{\vec{S}}\vec{v} + \widehat{P}_{\vec{S}^\perp}\vec{v} = \vec{v}_{\vec{S}} + \vec{v}_{\vec{S}^\perp} = \vec{v}.$$

We have used Eq (6.59). It follows that $\widehat{P}_{\vec{S}} + \widehat{P}_{\vec{S}^\perp} = \widehat{I\!I}$.

Q9(6) Find the eigenvectors of the two-dimensional projector in Eq. (9.43) corresponding to the eigenvalues 1 and 0.

SQ9(6) The vector $\vec{e}_3$ is an eigenvector of $\widehat{P}_{\vec{S}_{12}}$ corresponding to the eigenvalue 0 while $\vec{e}_1$ and $\vec{e}_2$ are eigenvectors of $\widehat{P}_{\vec{S}_{12}}$ corresponding to the eigenvalue 1. The eigenvalue 0 is nondegenerate while the eigenvalue 1 is degenerate with degeneracy 2. Any linear combination of $\vec{e}_1$ and $\vec{e}_2$ is also an eigenvector of $\widehat{P}_{\vec{S}_{12}}$ corresponding to the eigenvalue 1.

Q9(7) Verify Eq. (9.44).

SQ9(7) When $\langle \vec{e}_1 \mid \vec{e}_2 \rangle = 0$ we have $(\widehat{P}_{\vec{e}_1}\widehat{P}_{\vec{e}_2})\vec{u} = \widehat{P}_{\vec{e}_1}(\widehat{P}_{\vec{e}_2}\vec{u}) = \vec{0}$ since the projection $\widehat{P}_{\vec{e}_2}\vec{u}$ of $\vec{u}$ onto $\vec{e}_2$ has a zero projection onto $\vec{e}_1$. Next, suppose $\widehat{P}_{\vec{e}_1}\widehat{P}_{\vec{e}_2} = \widehat{0}$. Then for any $\vec{u}$ and $\vec{v}$ we have

$$\langle \widehat{P}_{\vec{e}_1}\vec{u} \mid \widehat{P}_{\vec{e}_2}\vec{v} \rangle = \langle \vec{u} \mid \widehat{P}_{\vec{e}_1}\widehat{P}_{\vec{e}_2}\vec{v} \rangle = 0.$$

Since

$$\widehat{P}_{\vec{e}_1}\vec{u} = u_1\vec{e}_1, \quad u_1 = \langle \vec{e}_1 \mid \vec{u} \rangle, \qquad \widehat{P}_{\vec{e}_2}\vec{v} = v_2\vec{e}_2, \quad v_2 = \langle \vec{e}_2 \mid \vec{v} \rangle,$$

we get

$$\langle \widehat{P}_{\vec{e}_1}\vec{u} \mid \widehat{P}_{\vec{e}_2}\vec{v} \rangle = \langle u_1\vec{e}_1 \mid v_2\vec{e}_2 \rangle = u_1 v_2 \langle \vec{e}_1 \mid \vec{e}_2 \rangle = 0.$$

It follows that $\langle \vec{e}_1 \mid \vec{e}_2 \rangle = 0$.

Q9(8) Prove Theorems 9.4.1(1), 9.4.1(2) and Corollary 9.4.1(1).

SQ9(8)[1] We start with Eq. (6.30) which implies

$$\langle \vec{v} \mid \widehat{A}\vec{w} \rangle = 0 \;\; \forall \vec{v} \;\; \Rightarrow \;\; \widehat{A}\vec{w} = \vec{0} \;\; \forall \vec{w} \;\; \Rightarrow \;\; \widehat{A} = \widehat{0}.$$

The proof of Theorem 9.4.1(1) is based on showing that

$$\langle \vec{u} \mid \widehat{A}\vec{u} \rangle = 0 \;\; \forall \vec{u} \;\; \Rightarrow \;\; \langle \vec{v} \mid \widehat{A}\vec{w} \rangle = 0 \;\; \forall \vec{v}, \vec{w}.$$

[1] Halmos p. 138.

This result can be established by expressing the scalar product $\langle \vec{v} \mid \widehat{A}\vec{w} \rangle$ as a sum of the values of the quadratic form generated by $\widehat{A}$ for different vectors. We can achieve this by expanding the quadratic form $\langle (\vec{v}+\vec{w}) \mid \widehat{A}(\vec{v}+\vec{w}) \rangle$, i.e.,

$$\langle (\vec{v}+\vec{w}) \mid \widehat{A}(\vec{v}+\vec{w}) \rangle = \langle \vec{v} \mid \widehat{A}\vec{v} \rangle + \langle \vec{w} \mid \widehat{A}\vec{w} \rangle + \langle \vec{v} \mid \widehat{A}\vec{w} \rangle + \langle \vec{w} \mid \widehat{A}\vec{v} \rangle.$$

Since $\widehat{A}$ is selfadjoint we have $\langle \vec{v} \mid \widehat{A}\vec{w} \rangle = \langle \widehat{A}\vec{v} \mid \vec{w} \rangle$. We also have $\langle \widehat{A}\vec{v} \mid \vec{w} \rangle = \langle \vec{w} \mid \widehat{A}\vec{v} \rangle$ since scalar product in $\vec{\mathbb{E}}^3$ is commutative. The above equation becomes

$$\langle (\vec{v}+\vec{w}) \mid \widehat{A}(\vec{v}+\vec{w}) \rangle = \langle \vec{v} \mid \widehat{A}\vec{v} \rangle + \langle \vec{w} \mid \widehat{A}\vec{w} \rangle + 2\langle \vec{v} \mid \widehat{A}\vec{w} \rangle.$$

We get

$$2\langle \vec{v} \mid \widehat{A}\vec{w} \rangle = \langle (\vec{v}+\vec{w}) \mid \widehat{A}(\vec{v}+\vec{w}) \rangle - \langle \vec{v} \mid \widehat{A}\vec{v} \rangle - \langle \vec{w} \mid \widehat{A}\vec{w} \rangle.$$

The condition $\langle \vec{u} \mid \widehat{A}\vec{u} \rangle = 0 \ \forall \vec{u}$ implies that the values of the quadratic form for $\vec{v}+\vec{w}$, $\vec{v}$ and $\vec{w}$ on the right hand side are all zero. The above equation implies $\langle \vec{v} \mid \widehat{A}\vec{w} \rangle = 0 \ \forall \vec{v}, \vec{w}$ which in turn implies $\widehat{A} = \widehat{0}$.
Theorem 9.4.1(2) follows from Theorem 9.4.1(1) since

$$\langle \vec{u} \mid \widehat{A}\vec{u} \rangle = \langle \vec{u} \mid \widehat{B}\vec{u} \rangle \ \forall \vec{u} \ \Rightarrow \ \langle \vec{u} \mid (\widehat{A}-\widehat{B})\vec{u} \rangle = 0 \ \forall \vec{u}$$
$$\Rightarrow \ \widehat{A} - \widehat{B} = \widehat{0}.$$

Corollary 9.4.1(1) follows immediately from Theorem 9.4.1(2) since

$$\langle \vec{u} \mid \widehat{A}\vec{u} \rangle = \langle \vec{u} \mid \vec{u} \rangle = \langle \vec{u} \mid \widehat{I\!I}\vec{u} \rangle \ \forall \vec{u} \ \Rightarrow \ \widehat{A} = \widehat{I\!I}.$$

Q9(9) Show that all projectors satisfy the sefladjointness condition in Eq. (9.35) and that they are also idempotent.

SQ9(9) From Definition 9.3.1(1) it is easily verified that any one-dimensional projector $\widehat{P}_{\vec{e}}$ satisfies the selfadjoint condition in Eq. (9.35), i.e., $\langle \vec{u} \mid \widehat{P}_{\vec{e}}\vec{v} \rangle = u_e v_e = \langle \widehat{P}_{\vec{e}}\vec{u} \mid \vec{v} \rangle$, where $u_e = \langle \vec{e} \mid \vec{u} \rangle$ and $v_e = \langle \vec{e} \mid \vec{v} \rangle$. Since

$$\widehat{P}_{\vec{e}}\big(\widehat{P}_{\vec{e}}\vec{u}\big) = \widehat{P}_{\vec{e}}\big(u_e\vec{e}\big) = u_e\widehat{P}_{\vec{e}}\vec{e} = u_e\vec{e} = \widehat{P}_{\vec{e}}\vec{u},$$

the operator is also idempotent. A two-dimensional projector $\widehat{P}_{\vec{S}}$ is expressible as the sum of two orthogonal one-dimensional projectors as shown in Eq. (9.43), i.e.,

$$\widehat{P}_{\vec{S}} = \widehat{P}_{\vec{e}_1} + \widehat{P}_{\vec{e}_2}, \quad \langle \vec{e}_1 \mid \vec{e}_2 \rangle = 0.$$

The selfadjointness property can be verified as follows:

$$\begin{aligned}\langle \vec{u} \mid \widehat{P}_{\vec{S}} \vec{v} \rangle &= \langle \vec{u} \mid (\widehat{P}_{\vec{e}_1} + \widehat{P}_{\vec{e}_2})\, \vec{v} \rangle = \langle \vec{u} \mid \widehat{P}_{\vec{e}_1} \vec{v} \rangle + \langle \vec{u} \mid \widehat{P}_{\vec{e}_2} \vec{v} \rangle \\ &= \langle \widehat{P}_{\vec{e}_1} \vec{u} \mid \vec{v} \rangle + \langle \widehat{P}_{\vec{e}_2} \vec{u} \mid \vec{v} \rangle = \langle (\widehat{P}_{\vec{e}_1} + \widehat{P}_{\vec{e}_2})\, \vec{u} \mid \vec{v} \rangle \\ &= \langle \widehat{P}_{\vec{S}} \vec{u} \mid \vec{v} \rangle.\end{aligned}$$

Since $\widehat{P}_{\vec{e}_1}$ and $\widehat{P}_{\vec{e}_2}$ are idempotent and orthogonal we have

$$\left(\widehat{P}_{\vec{e}_1} + \widehat{P}_{\vec{e}_2}\right)^2 = \widehat{P}^2_{\vec{e}_1} + \widehat{P}^2_{\vec{e}_2} = \widehat{P}_{\vec{e}_1} + \widehat{P}_{\vec{e}_2},$$

showing that $\widehat{P}_{\vec{e}_1} + \widehat{P}_{\vec{e}_2}$ is idempotent.

Q9(10) Show that selfadjoint operators are represented by selfadjoint matrices.

SQ9(10) The matrix representation of $\widehat{A}^\dagger$ is the adjoint matrix to the matrix representation to $\widehat{A}$, as shown in SQ8(6). For a selfadjoint operator we have $\widehat{A}^\dagger = \widehat{A}$ which implies that matrix representation of $\widehat{A}^\dagger$ is the same as matrix representation of $\widehat{A}$, i.e., the matrix is selfadjoint.

Q9(11) Demonstrate the inequality in Eq. (9.53) with $\vec{u} = u_y \vec{j}$ and $\vec{v} = v_x \vec{i}$. Verify Eq. (9.56).

SQ9(11) For the orthogonal operator $\widehat{R}_z(\theta_z)$ in Eq. (9.2) and for any two vectors $\vec{u}$ and $\vec{v}$ lying in the x-y plane, i.e., $\vec{u} = u_x \vec{i} + u_y \vec{j}$ and $\vec{v} = v_x \vec{i} + v_y \vec{j}$, we have, in accordance with Eq. (9.52),

$$\begin{aligned}\widehat{R}_z(\theta_z)\vec{u} &= (u_x \cos\theta_z - u_y \sin\theta_z)\,\vec{i} + (u_x \sin\theta_z + u_y \cos\theta_z)\,\vec{j}, \\ \widehat{R}_z(\theta_z)\vec{v} &= (v_x \cos\theta_z - v_y \sin\theta_z)\,\vec{i} + (v_x \sin\theta_z + v_y \cos\theta_z)\,\vec{j}.\end{aligned}$$

Then

$$\begin{aligned}\langle \vec{u} \mid \widehat{R}_z(\theta_z)\vec{v} \rangle &= u_x(v_x \cos\theta_z - v_y \sin\theta_z) \\ &\quad + u_y(v_x \sin\theta_z + v_y \cos\theta_z). \\ \langle \widehat{R}_z(\theta_z)\vec{u} \mid \vec{v} \rangle &= (u_x \cos\theta_z - u_y \sin\theta_z)v_x \\ &\quad + (u_x \sin\theta_z + u_y \cos\theta_z)v_y.\end{aligned}$$

When $\vec{u} = u_y\, \vec{j}$ with $u_x = 0$ and $\vec{v} = v_x\, \vec{i}$ with $v_y = 0$ we get

$$\langle \vec{u} \mid \widehat{R}_z(\theta_z)\vec{v} \rangle = u_y v_x \, \sin\theta_z \neq \langle \widehat{R}_z(\theta_z)\vec{u} \mid \vec{v} \rangle = -u_y v_x \, \sin\theta_z.$$

Hence the operator $\widehat{R}_z(\theta_z)$ is not selfadjoint. The corresponding matrix $\boldsymbol{R}_z(\theta_z)$ in Eq. (7.148) is also not selfadjoint.

When acting on $\vec{u} = u_x\vec{i} + u_y\vec{j}$ the adjoint operator is given by Eqs. (9.55), i.e.,

$$\widehat{R}_z^\dagger(\theta_z)\vec{u} = u'_x\vec{i} + u'_y\vec{j},$$

where

$$u'_x = u_x \cos\theta_z + u_y \sin\theta_z, \quad u'_y = -u_x \sin\theta_z + u_y \cos\theta_z.$$

This operator is seen to be different from $\widehat{R}_z(\theta_z)$. We have

$\langle \widehat{R}_z^\dagger(\theta_z)\vec{u} \mid \vec{v}\,\rangle = (u_x \cos\theta_z + u_y \sin\theta_z)v_x + (-u_x \sin\theta_z + u_y \cos\theta_z)v_y.$

This agrees with the expression for $\langle \vec{u} \mid \widehat{R}_z(\theta_z)\vec{v}\,\rangle$ obtained earlier.

Q9(12) Prove Eq. (9.70).

SQ9(12) Let

$$\vec{e}\,'_{21} = \alpha_1\vec{e}_{21} + \alpha_2\vec{e}_{22}, \quad \vec{e}\,'_{22} = \beta_1\vec{e}_{21} + \beta_2\vec{e}_{22}.$$

Since $\vec{e}\,'_{21}$ and $\vec{e}\,'_{22}$ are orthonormal the coefficients must satisfy

$$\alpha_1^2 + \alpha_2^2 = 1, \quad \beta_1^2 + \beta_2^2 = 1, \quad \text{and} \quad \alpha_1\beta_1 + \alpha_2\beta_2 = 0.$$

We can also express $\vec{e}_{21}$ and $\vec{e}_{21}$ in terms of $\vec{e}\,'_{21}$ and $\vec{e}\,'_{21}$, i.e.,

$$\vec{e}_{21} = \frac{\beta_2\vec{e}\,'_{21} - \alpha_2\vec{e}\,'_{22}}{\beta_2\alpha_1 - \alpha_2\beta_1}, \quad \vec{e}_{22} = \frac{\beta_1\vec{e}\,'_{21} - \alpha_1\vec{e}\,'_{22}}{\beta_1\alpha_2 - \alpha_1\beta_2}.$$

Since $\vec{e}_{21}$ and $\vec{e}_{22}$ are

(1) orthogonal the coefficients must satisfy

$$\langle \beta_2\vec{e}\,'_{21} - \alpha_2\vec{e}\,'_{22} \mid \beta_1\vec{e}\,'_{21} - \alpha_1\vec{e}\,'_{22}\rangle = 0 \;\Rightarrow\; \alpha_1\alpha_2 + \beta_1\beta_2 = 0,$$

(2) normalised the coefficients must satisfy

$$\beta_2^2 + \alpha_2^2 = (\beta_2\alpha_1 - \alpha_2\beta_1)^2 = \alpha_1^2 + \beta_1^2.$$

Adding β_1^2 and α_1^2 on both sides we get

$$\beta_2^2 + \beta_1^2 + \alpha_2^2 + \alpha_1^2 = 2(\alpha_1^2 + \beta_1^2).$$

Since $\beta_2^2 + \beta_1^2 = \alpha_2^2 + \alpha_1^2 = 1$ we get $\alpha_1^2 + \beta_1^2 = 1$ which in turn implies $\alpha_2^2 + \beta_2^2 = 1$.

For any given $\vec{u}$ we have

$$\big(\widehat{P}_{\vec{e}_{21}} + \widehat{P}_{\vec{e}_{22}}\big)\vec{u} = u_1\,\vec{e}_{21} + u_2\,\vec{e}_{22},$$

where $u_1 = \langle \vec{e}_{21} \mid \vec{u}\,\rangle$ and $u_2 = \langle \vec{e}_{22} \mid \vec{u}\,\rangle$ and

$$\langle \vec{u} \mid \big(\widehat{P}_{\vec{e}_{21}} + \widehat{P}_{\vec{e}_{22}}\big)\vec{u}\,\rangle = u_1^2 + u_2^2.$$

For the projectors generated by $\vec{e}\,'_{21}$ and $\vec{e}\,'_{22}$ we have

$$\widehat{P}_{\vec{e}\,'_{21}}\vec{u} = \langle \vec{e}\,'_{21} \mid \vec{u}\,\rangle\,\vec{e}\,'_{21} = \big(\alpha_1 u_1 + \alpha_2 u_2\big)\big(\alpha_1\vec{e}_{21} + \alpha_2\vec{e}_{22}\big),$$
$$\widehat{P}_{\vec{e}\,'_{22}}\vec{u} = \langle \vec{e}\,'_{22} \mid \vec{u}\,\rangle\,\vec{e}\,'_{22} = \big(\beta_1 u_1 + \beta_2 u_2\big)\big(\beta_1\vec{e}_{21} + \beta_2\vec{e}_{22}\big).$$

We get

$$\begin{aligned}
\langle \vec{u} \mid \widehat{P}_{\vec{e}\,'_{21}}\vec{u}\,\rangle &= \big(\alpha_1 u_1 + \alpha_2 u_2\big)\,\langle \vec{u} \mid \big(\alpha_1\vec{e}_{21} + \alpha_2\vec{e}_{22}\big)\rangle \\
&= \big(\alpha_1 u_1 + \alpha_2 u_2\big)\big(\alpha_1 u_1 + \alpha_2 u_2\big) \\
&= \alpha_1^2 u_1^2 + 2\alpha_1\alpha_2 u_1 u_2 + \alpha_2^2 u_2^2, \\
\langle \vec{u} \mid \widehat{P}_{\vec{e}\,'_{22}}\vec{u}\,\rangle &= \big(\beta_1 u_1 + \beta_2 u_2\big)\langle \vec{u} \mid \big(\beta_1\vec{e}_{21} + \beta_2\vec{e}_{22}\big)\rangle \\
&= \big(\beta_1 u_1 + \beta_2 u_2\big)\big(\beta_1 u_1 + \beta_2 u_2\big) \\
&= \beta_1^2 u_1^2 + 2\beta_1\beta_2 u_1 u_2 + \beta_2^2 u_2^2.
\end{aligned}$$

Adding the above equations we get, for all $\vec{u}$,

$$\begin{aligned}
&\langle \vec{u} \mid \big(\widehat{P}_{\vec{e}\,'_{21}} + \widehat{P}_{\vec{e}\,'_{22}}\big)\vec{u}\,\rangle \\
&= \big(\alpha_1^2 + \beta_1^2\big)\,u_1^2 + \big(\alpha_2^2 + \beta_2^2\big)\,u_2^2 \quad \text{due to } (\alpha_1\alpha_2 + \beta_1\beta_2) = 0 \\
&= u_1^2 + u_2^2 \quad \text{due to } \alpha_1^2 + \beta_1^2 = 1 \text{ and } \alpha_2^2 + \beta_2^2 = 1 \\
&= \langle \vec{u} \mid \big(\widehat{P}_{\vec{e}_{21}} + \widehat{P}_{\vec{e}_{22}}\big)\vec{u}\,\rangle \\
&\Rightarrow \widehat{P}_{\vec{e}\,'_{21}} + \widehat{P}_{\vec{e}\,'_{22}} = \widehat{P}_{\vec{e}_{21}} + \widehat{P}_{\vec{e}_{22}}.
\end{aligned}$$

The present question is relevant when the eigenvalue a_2 of a selfadjoint operator $\widehat{A}$ is degenerate with degeneracy 2 corresponding to two orthonormal eigenvectors $\vec{e}_{21}$, $\vec{e}_{22}$. The corresponding eigensubspace is two-dimensional spanned by $\vec{e}_{21}, \vec{e}_{22}$. The projector onto this two-dimensional subspace is given by Eq. (9.43), i.e., equal to the sum of two projectors $\widehat{P}_{\vec{e}_{21}}$ and $\widehat{P}_{\vec{e}_{22}}$, i.e.,

$$\begin{aligned}
\widehat{P}^{\widehat{A}}(a_2) &= \widehat{P}_{\vec{e}_{21}} + \widehat{P}_{\vec{e}_{22}} = |\vec{e}_{21}\rangle\langle\vec{e}_{21}| + |\vec{e}_{22}\rangle\langle\vec{e}_{22}| \\
&= \widehat{P}_{\vec{e}\,'_{21}} + \widehat{P}_{\vec{e}\,'_{22}} = |\vec{e}\,'_{21}\rangle\langle\vec{e}\,'_{21}| + |\vec{e}\,'_{22}\rangle\langle\vec{e}\,'_{22}|.
\end{aligned}$$

which is Eq. (9.70).

Q9(13) Show that the selfadjoint operator $\widehat{A}$ in Eq. (9.71) admits a_ℓ as its eigenvalues with the unit vectors $\vec{e}_\ell$ as the corresponding eigenvectors.

SQ9(13) Given the operator $\widehat{A}$ in Eq. (9.71) we have

$$\widehat{A}\vec{e}_\ell = \Big(\sum_{\ell'}^{3} a_{\ell'}\,\widehat{P}_{\vec{e}_{\ell'}}\Big)\vec{e}_\ell = \sum_{\ell'}^{3} a_{\ell'}\,\widehat{P}_{\vec{e}_{\ell'}}\vec{e}_\ell = \sum_{\ell'}^{3} a_{\ell'}\,\delta_{\ell'\ell}\vec{e}_\ell = a_\ell\vec{e}_\ell.$$

This shows that $\vec{e}_\ell$ is an eigenvector of $\widehat{A}$ corresponding to eigenvalue a_ℓ.

Q9(14) Show that the products of an operator with its adjoint, i.e., $\widehat{A}^\dagger\widehat{A}$ and $\widehat{A}\widehat{A}^\dagger$, are selfadjoint and positive.

SQ9(14) Using Eqs. (8.40) and (8.41) we get

$$(\widehat{A}^\dagger\widehat{A})^\dagger = \widehat{A}^\dagger\widehat{A}^{\dagger\dagger} = (\widehat{A}^\dagger\widehat{A}) \quad \text{and} \quad (\widehat{A}\widehat{A}^\dagger)^\dagger = \widehat{A}^{\dagger\dagger}\widehat{A}^\dagger = (\widehat{A}\widehat{A}^\dagger).$$

These results imply that $\widehat{A}^\dagger\widehat{A}$ and $\widehat{A}\widehat{A}^\dagger$ are selfadjoint. Moreover we have

$$\langle \vec{v} \mid (\widehat{A}\widehat{A}^\dagger)\vec{v}\,\rangle = \langle \widehat{A}^\dagger\vec{v} \mid \widehat{A}^\dagger\vec{v}\,\rangle = ||\,\widehat{A}^\dagger\vec{v}\,||^2 \geq 0,$$
$$\langle \vec{v} \mid (\widehat{A}^\dagger\widehat{A})\vec{v}\,\rangle = \langle \widehat{A}\vec{v} \mid \widehat{A}\vec{v}\,\rangle = ||\,\widehat{A}\vec{v}\,||^2 \geq 0.$$

These two equations hold for all $\vec{v}$, including the eigenvectors of $\widehat{A}^\dagger\widehat{A}$ and $\widehat{A}\widehat{A}^\dagger$. It follows that the product operators do not have negative eigenvalues, i.e., both product operators are positive.

Q9(15) Show that the expression of the inverse in Eq. (9.86) satisfy the condition $\widehat{A}\widehat{A}^{-1} = \widehat{I\!I}$.

SQ9(15) Taking the expressions for $\widehat{A}$ in Eq. (9.79) and $\widehat{A}^{-1}$ in Eq. (9.86) we get

$$\begin{aligned}\widehat{A}\widehat{A}^{-1} &= \Big(\sum_{m'} a_{m'}\,\widehat{P}^{\hat{A}}(a_{m'})\Big)\Big(\sum_{m} a_m^{-1}\,\widehat{P}^{\hat{A}}(a_m)\Big)\\ &= \sum_{m',m} a_{m'}a_m^{-1}\,\widehat{P}^{\hat{A}}(a_{m'})\widehat{P}^{\hat{A}}(a_m)\\ &= \sum_{m',m} a_{m'}a_m^{-1}\,\widehat{P}^{\hat{A}}(a_{m'})\delta_{m'm}\\ &= \sum_{m} \widehat{P}^{\hat{A}}(a_m) = \widehat{I\!I}.\end{aligned}$$

We have used the result $\widehat{P}^{\hat{A}}(a_{m'})\,\widehat{P}^{\hat{A}}(a_m) = \widehat{P}^{\hat{A}}(a_m)\,\delta_{m'm}$ and the spectral decomposition of the identity in Eq. (9.80).

Q9(16) The square root $\sqrt{\widehat{A}}$ of a positive operator $\widehat{A}$ is defined by Eq. (9.87). Show that $\left(\sqrt{\widehat{A}}\,\right)^2 = \widehat{A}$.

SQ9(16) From Eq. (9.87) we get

$$\begin{aligned}\left(\sqrt{\widehat{A}}\right)^2 &= \Big(\sum_m \sqrt{a_m}\,\widehat{P}^{\hat{A}}(a_m)\Big)\Big(\sum_{m'} \sqrt{a_{m'}}\,\widehat{P}^{\hat{A}}(a_{m'})\Big)\\ &= \sum_{m,m'} \sqrt{a_m}\,\sqrt{a_{m'}}\,\widehat{P}^{\hat{A}}(a_m)\,\delta_{mm'}\\ &= \sum_m a_m\,\widehat{P}^{\hat{A}}(a_m) = \widehat{A}.\end{aligned}$$

We have used the result $\widehat{P}^{\hat{A}}(a_m)\,\widehat{P}^{\hat{A}}(a_{m'}) = \widehat{P}^{\hat{A}}(a_m)\,\delta_{mm'}$ and the spectral decomposition of the $\widehat{A}$ in Eq. (9.79).

Q9(17) Show that the quadratic form $\mathcal{Q}(\widehat{A}, \vec{u}\,)$ generated by a positive selfadjoint operator in any $\vec{u}$ is non-negative.

SQ9(17) Since the quadratic form $\mathcal{Q}(\widehat{A}, \vec{u}\,)$ is given by $\langle \vec{u} \mid \widehat{A}\,\vec{u}\,\rangle$ we get, for a positive selfadjoint operator $\widehat{A}$,

$$\begin{aligned}\mathcal{Q}(\widehat{A}, \vec{u}\,) &= \langle \vec{u} \mid \sum_m a_m\,\widehat{P}^{\hat{A}}(a_m)\,\vec{u}\,\rangle, \quad a_m \geq 0\\ &= \sum_m a_m\,\langle \vec{u} \mid \widehat{P}^{\hat{A}}(a_m)\,\vec{u}\,\rangle\\ &\geq 0,\end{aligned}$$

since the eigenvalues a_m of a positive selfadjoint operator $\widehat{A}$ is non-negative by definition 9.4.4(2) and

$$\begin{aligned}\langle \vec{u} \mid \widehat{P}^{\hat{A}}(a_m)\,\vec{u}\,\rangle &= \langle \vec{u} \mid \left(\widehat{P}^{\hat{A}}(a_m)\right)^2 \vec{u}\,\rangle\\ &= \langle \widehat{P}^{\hat{A}}(a_m)\vec{u} \mid \widehat{P}^{\hat{A}}(a_m)\,\vec{u}\,\rangle\\ &= ||\,\widehat{P}^{\hat{A}}(a_m)\vec{u}\,||^2 \geq 0.\end{aligned}$$

Chapter 10

Probability, Selfadjoint Operators, Unit Vectors and the Need for Complexness

Q10(1) In the model theory in §10.2.1 the observable A has three possible values, i.e., a_1, a_2, and a_3. Find the probabilities for the value a_1 when the state vector $\vec{u}$ is one of the following unit vectors:

$$\vec{u}_1 = \vec{e}_1, \quad \vec{u}_2 = \vec{e}_2, \quad \text{and} \quad \vec{u}_3 = \frac{1}{\sqrt{3}}\vec{e}_1 + \sqrt{\frac{2}{3}}\,\vec{e}_3.$$

What are the expectation values $\mathcal{E}(\widehat{A}, \vec{u}\,)$ and the uncertainties $\Delta(\widehat{A}, \vec{u}\,)$ in these cases?

SQ10(1)

(1) For the first state vector $\vec{u}_1 = \vec{e}_1$ the probability of getting the value a_1 on a measurement of A is 1 with expectation value $\mathcal{E}(\widehat{A}, \vec{u}_1) = a_1$ and zero uncertainty.

(2) For the second state vector $\vec{u}_2 = \vec{e}_2$ the probability of getting the value a_1 on a measurement of A is 0 with expectation value $\mathcal{E}(\widehat{A}, \vec{u}_2) = a_2$ and zero uncertainty.

(3) For the third state vector $\vec{u}_3 = (\vec{e}_1 + \sqrt{2}\,\vec{e}_3)/\sqrt{3}$ the probability of getting the value a_1 on a measurement of A is $1/3$ with

Quantum Mechanics: Problems and Solutions
K. Kong Wan

ISBN 978-981-4800-72-3 (Paperback), 978-0-429-29647-5 (eBook)
www.jennystanford.com

expectation value

$$\mathcal{E}(\widehat{A}, \vec{u}_3) = \frac{1}{3}(a_1 + 2a_3).$$

The variant can be calculated in accordance with Eq. (3.33), i.e., we have

$$\begin{aligned}\text{Var}(A, \vec{u}_3) &= \left(a_1 - \mathcal{E}(\widehat{A}, \vec{u}_3)\right)^2 \frac{1}{3} + \left(a_3 - \mathcal{E}(\widehat{A}, \vec{u}_3)\right)^2 \frac{2}{3} \\ &= \frac{4}{27}(a_1 - a_3)^2 + \frac{2}{27}(-a_1 + a_3)^2 \\ &= \frac{2}{9}(a_1 - a_3)^2.\end{aligned}$$

The uncertainty is then given by

$$\Delta(A, \vec{u}_3) = \sqrt{\text{Var}(A, \vec{u}_3)} = \frac{\sqrt{2}}{3}\,|\,a_1 - a_3\,|.$$

The variant can also be calculated in accordance with Eq. (3.36), i.e., we have

$$\begin{aligned}\text{Var}(A, \vec{u}_3) &= \left(a_1^2 \frac{1}{3} + a_3^2 \frac{2}{3}\right) - \mathcal{E}(\widehat{A}, \vec{u})^2 \\ &= \frac{1}{3}(a_1^2 + 2a_3^2) - \frac{1}{9}(a_1 + 2a_3)^2 \\ &= \frac{2}{9}(a_1 - a_3)^2.\end{aligned}$$

Q10(2) Show that $\widehat{A}$ and $\widehat{A}'$ in Eq. (10.16) commute.

SQ10(2) Writing

$$\widehat{A} = \sum_{\ell=1}^{3} a_\ell \widehat{P}_{\vec{e}_\ell} \quad \text{and} \quad \widehat{A}' = \sum_{\ell'=1}^{3} a'_{\ell'} \widehat{P}_{\vec{e}_{\ell'}}.$$

we get

$$\widehat{A}\widehat{A}' = \sum_{\ell,\ell'=1}^{3} a_\ell a'_{\ell'} \widehat{P}_{a_\ell} \widehat{P}_{a_{\ell'}} = \sum_{\ell,\ell'=1}^{3} a_\ell a'_{\ell'} \widehat{P}_{a_\ell} \delta_{\ell\ell'} = \sum_{\ell=1}^{3} a_\ell a'_\ell \widehat{P}_{a_\ell}.$$

$$\widehat{A}'\widehat{A} = \sum_{\ell',\ell=1}^{3} a'_{\ell'} a_\ell \widehat{P}_{a_{\ell'}} \widehat{P}_{a_\ell} = \sum_{\ell,\ell'=1}^{3} a'_{\ell'} a_\ell \widehat{P}_{a_\ell} \delta_{\ell\ell'} = \sum_{\ell=1}^{3} a'_\ell a_\ell \widehat{P}_{a_\ell}.$$

It follows that $\widehat{A}\widehat{A}' - \widehat{A}'\widehat{A} = \widehat{0}$, i.e., $\widehat{A}$ and $\widehat{A}'$ commute.

Q10(3) Prove Eq. (10.31).

SQ10(3) From Eq. (10.30) we get

$$|z^*| = \sqrt{z^{**}z^*} = \sqrt{zz^*} = \sqrt{z^*z} = |z|.$$

More directly we have

$$|z^*| = \sqrt{a^2 + b^2} = |z|.$$

For $z = a + ib$ and $w = c + id$. We have

$$|z|\,|w| = \sqrt{(a^2 + b^2)}\sqrt{(c^2 + d^2)} = \sqrt{(a^2 + b^2)(c^2 + d^2)}.$$

From Eq. (10.29) on product of two complex numbers we get

$$zw = (ac - bd) + i\,(ad + bc).$$

The norm of zw is given by Eq. (10.30), i.e. we get

$$\begin{aligned}|zw| &= \sqrt{(ac - bd)^2 + (ad + bc)^2} \\ &= \sqrt{(ac)^2 + (bd)^2 + (ad)^2 + (bc)^2} \\ &= \sqrt{(a^2 + b^2)(c^2 + d^2)} \\ &= |z|\,|w|.\end{aligned}$$

Q10(4) Let θ, a and b be real numbers. For the complex numbers $z = e^{i\theta}$ and $z = e^{i\theta}(a + ib)$ show that

$$|e^{i\theta}| = 1 \quad \text{and} \quad |z|^2 = a^2 + b^2.$$

SQ10(4) Using the Euler's formula given in Eq. (10.26) we get

$$\begin{aligned}e^{i\theta} &= \cos\theta + i\sin\theta \;\Rightarrow\; \left(e^{i\theta}\right)^* = \cos\theta - i\sin\theta \\ \Rightarrow\; |e^{i\theta}|^2 &= \left(e^{i\theta}\right)^* e^{i\theta} = (\cos\theta - i\sin\theta)(\cos\theta + i\sin\theta) = 1 \\ \Rightarrow\quad |z|^2 &= |e^{i\theta}|^2\,|(a + ib)|^2 = a^2 + b^2.\end{aligned}$$

We have used Eqs. (10.30) and (10.31) and the formula $\sin^2\theta + \cos^2\theta = 1$.

Chapter 11

Complex Vectors

Q11(1) Prove Eqs. (11.17), (11.18) and (11.19) in $\vec{\mathbb{E}}^{3c}$.

SQ11(1) To prove Eq. (11.17) we take the scalar product of $\vec{e}_\ell^{\,c}$ and $\vec{\zeta}$ to get

$$\langle \vec{e}_\ell^{\,c} \mid \vec{\zeta}\,\rangle^c = \langle \vec{e}_\ell^{\,c} \mid \sum_{\ell'=1}^{3} \zeta_{\ell'} \vec{e}_{\ell'}^{\,c}\,\rangle^c = \sum_{\ell'=1}^{3} \zeta_{\ell'} \langle \vec{e}_\ell^{\,c} \mid \vec{e}_{\ell'}^{\,c}\,\rangle^c = \zeta_\ell,$$

since $\langle \vec{e}_\ell^{\,c} \mid \vec{e}_{\ell'}^{\,c}\,\rangle^c = \delta_{\ell\ell'}$. A comparison with Eq. (11.12) shows that $\zeta_\ell = u_\ell + iv_\ell$. To prove Eq. (11.19) we have

$$\begin{aligned}\langle \vec{\zeta}_1 \mid \vec{\zeta}_2 \rangle^c &= \langle \sum_{\ell=1}^{3} \zeta_{1\ell} \vec{e}_\ell^{\,c} \mid \sum_{\ell'=1}^{3} \zeta_{2\ell'} \vec{e}_{\ell'}^{\,c}\,\rangle^c \\ &= \sum_{\ell,\ell'=1}^{3} \zeta_{1\ell}^* \zeta_{2\ell'} \langle \vec{e}_\ell^{\,c} \mid \vec{e}_{\ell'}^{\,c}\,\rangle^c = \sum_{\ell=1}^{3} \zeta_{1\ell}^* \zeta_{2\ell}.\end{aligned}$$

Equation (11.18) is just a special case of Eq. (11.19).

Q11(2) Let

$$\vec{\epsilon}_1 = \frac{1}{\sqrt{2}}\left(\vec{i}^{\,c} + i\vec{j}^{\,c}\right), \quad \vec{\epsilon}_2 = \frac{1}{\sqrt{2}}\left(\vec{i}^{\,c} - i\vec{j}^{\,c}\right), \quad \vec{\epsilon}_3 = \vec{k}^{\,c}.$$

(a) Write down the matrix representations of the above vectors in the orthonormal basis $\{\vec{i}^{\,c}, \vec{j}^{\,c}, \vec{k}^{\,c}\}$.

Quantum Mechanics: Problems and Solutions
K. Kong Wan

ISBN 978-981-4800-72-3 (Paperback), 978-0-429-29647-5 (eBook)
www.jennystanford.com

(b) Show that $\vec{\epsilon}_\ell$, $\ell = 1, 2, 3$ above form an orthonormal basis for $\vec{I\!E}^{3c}$.

(c) Show that an arbitrary vector $\vec{\zeta}$ can be expressed as

$$\vec{\zeta} = \sum_{\ell=1}^{3} \zeta_\ell \, \vec{\epsilon}_\ell, \quad \zeta_\ell = \langle \vec{\epsilon}_\ell \mid \vec{\zeta} \,\rangle^c.$$

SQ11(2)(a) The required matrix representations are given respectively by

$$\frac{1}{\sqrt{2}}\begin{pmatrix} 1 \\ i \\ 0 \end{pmatrix}, \quad \frac{1}{\sqrt{2}}\begin{pmatrix} 1 \\ -i \\ 0 \end{pmatrix}, \quad \begin{pmatrix} 0 \\ 0 \\ 1 \end{pmatrix}.$$

SQ11(2)(b) Since orthogonality implies linear independence all we need is to show that the three vectors are orthonormal.[1]

(1) The vector $\vec{\epsilon}_3$ is clearly normalised and orthogonal to $\vec{\epsilon}_1$ and $\vec{\epsilon}_2$.

(2) The vectors $\vec{\epsilon}_1$ and $\vec{\epsilon}_2$ are normalised since

$$\begin{aligned} \langle \vec{\epsilon}_1 \mid \vec{\epsilon}_1 \rangle^c &= \frac{1}{2} \langle \vec{i}^c + i\,\vec{j}^c \mid \vec{i}^c + i\,\vec{j}^c \,\rangle^c \\ &= \frac{1}{2} \langle \vec{i}^c \mid \vec{i}^c \,\rangle^c + i^* i \frac{1}{2} \langle \vec{j}^c \mid \vec{j}^c \,\rangle^c = 1. \end{aligned}$$

$$\begin{aligned} \langle \vec{\epsilon}_2 \mid \vec{\epsilon}_2 \rangle^c &= \frac{1}{2} \langle \vec{i}^c - i\,\vec{j}^c \mid \vec{i}^c - i\,\vec{j}^c \,\rangle^c \\ &= \frac{1}{2} \langle \vec{i}^c \mid \vec{i}^c \,\rangle^c + i^* i \frac{1}{2} \langle \vec{j}^c \mid \vec{j}^c \,\rangle^c = 1. \end{aligned}$$

(3) The vectors $\vec{\epsilon}_1$ and $\vec{\epsilon}_2$ are orthogonal since

$$\begin{aligned} \langle \vec{\epsilon}_1 \mid \vec{\epsilon}_2 \rangle^c &= \frac{1}{2} \langle \vec{i}^c + i\,\vec{j}^c \mid \vec{i}^c - i\,\vec{j}^c \rangle^c \\ &= \frac{1}{2} \langle \vec{i}^c \mid \vec{i}^c \,\rangle^c - i^* i \frac{1}{2} \langle \vec{j}^c \mid \vec{j}^c \,\rangle^c = 0. \end{aligned}$$

It follows that $\vec{\epsilon}_1$, $\vec{\epsilon}_2$ and $\vec{\epsilon}_3$ form an orthonormal basis in $\vec{I\!E}^{3c}$, since we know that $\vec{I\!E}^{3c}$ is three-dimensional.

---

[1] *See* Q6(4) and SQ6(4) or Q12(8) and SQ12(8).

SQ11(2)(c) The proof is the same as that in SQ11(1). Since $\vec{\epsilon}_1$, $\vec{\epsilon}_2$ and $\vec{\epsilon}_3$ form an orthonormal basis we can express any vector $\vec{\zeta}$ as a linear combination of $\vec{\epsilon}_1$, $\vec{\epsilon}_2$ and $\vec{\epsilon}_3$, i.e., we have $\vec{\zeta} = \sum_{\ell=1}^{3} \zeta_\ell \vec{\epsilon}_\ell$. Taking the scalar product of $\vec{\epsilon}_\ell$ and $\vec{\zeta}$ we get

$$\langle \vec{\epsilon}_\ell \mid \vec{\zeta}\,\rangle^c = \langle \vec{\epsilon}_\ell \mid \sum_{\ell'=1}^{3} \zeta_{\ell'} \vec{\epsilon}_{\ell'}\,\rangle^c = \sum_{\ell'=1}^{3} \zeta_{\ell'} \langle \vec{\epsilon}_\ell \mid \vec{\epsilon}_{\ell'}\,\rangle^c = \zeta_\ell.$$

Q11(3) Show that Eqs. (11.18) and (11.19) hold in a general complex orthonormal basis.

SQ11(3) The proof follows that of SQ11(1).

Chapter 12

N-Dimensional Complex Vectors

Q12(1) Prove Eq. (12.6).

SQ12(1) Evaluating $\langle \vec{\zeta} + \vec{\eta} \mid \vec{\zeta} + \vec{\eta} \rangle$ and $\langle \vec{\zeta} + i\vec{\eta} \mid \vec{\zeta} + i\vec{\eta} \rangle$ we get

$$\begin{aligned} \langle \vec{\zeta} + \vec{\eta} \mid \vec{\zeta} + \vec{\eta} \rangle &= \langle \vec{\zeta} \mid \vec{\zeta} \rangle + \langle \vec{\eta} \mid \vec{\eta} \rangle + \langle \vec{\zeta} \mid \vec{\eta} \rangle + \langle \vec{\eta} \mid \vec{\zeta} \rangle. \\ \langle \vec{\zeta} + i\vec{\eta} \mid \vec{\zeta} + i\vec{\eta} \rangle &= \langle \vec{\zeta} \mid \vec{\zeta} \rangle + \langle \vec{\eta} \mid \vec{\eta} \rangle + i\langle \vec{\zeta} \mid \vec{\eta} \rangle - i\langle \vec{\eta} \mid \vec{\zeta} \rangle. \\ \Rightarrow \quad \langle \vec{\zeta} + \vec{\eta} \mid \vec{\zeta} + \vec{\eta} \rangle & - i \, \langle \vec{\zeta} + i\vec{\eta} \mid \vec{\zeta} + i\vec{\eta} \rangle \\ &= (1-i)\big(\langle \vec{\zeta} \mid \vec{\zeta} \rangle + \langle \vec{\eta} \mid \vec{\eta} \rangle\big) + 2\langle \vec{\zeta} \mid \vec{\eta} \rangle. \end{aligned}$$

Equation (12.6) follows immediately.

Q12(2) Show that

$$\begin{aligned} \langle \vec{\zeta} \mid \vec{\eta} \rangle = \frac{1}{4}\Big\{ & ||\vec{\zeta} + \vec{\eta}||^2 - ||\vec{\zeta} - \vec{\eta}||^2 \\ & - i\,\big(||\vec{\zeta} + i\vec{\eta}||^2 - ||\vec{\zeta} - i\vec{\eta}||^2 \big) \Big\}. \end{aligned}$$

SQ12(2) To prove the above equation we begin by evaluating

$$||\vec{\zeta} + \vec{\eta}||^2 - ||\vec{\zeta} - \vec{\eta}||^2 \quad \text{and} \quad ||\vec{\zeta} + i\vec{\eta}||^2 - ||\vec{\zeta} - i\vec{\eta}||^2.$$

The calculations run as follows

$$\begin{aligned} ||\vec{\zeta} + \vec{\eta}||^2 - ||\vec{\zeta} - \vec{\eta}||^2 &= \langle \vec{\zeta} + \vec{\eta} \mid \vec{\zeta} + \vec{\eta} \rangle - \langle \vec{\zeta} - \vec{\eta} \mid \vec{\zeta} - \vec{\eta} \rangle \\ &= 2\,\langle \vec{\zeta} \mid \vec{\eta} \rangle + 2\,\langle \vec{\eta} \mid \vec{\zeta} \rangle. \\ ||\vec{\zeta} + i\vec{\eta}||^2 - ||\vec{\zeta} - i\vec{\eta}||^2 &= 2i\,\langle \vec{\zeta} \mid \vec{\eta} \rangle - 2i\,\langle \vec{\eta} \mid \vec{\zeta} \rangle. \end{aligned}$$

Quantum Mechanics: Problems and Solutions
K. Kong Wan

ISBN 978-981-4800-72-3 (Paperback), 978-0-429-29647-5 (eBook)
www.jennystanford.com

The desired result follows, i.e., we have

$$\|\vec{\zeta}+\vec{\eta}\|^2-\|\vec{\zeta}-\vec{\eta}\|^2-i\left(\|\vec{\zeta}+i\vec{\eta}\|^2+i\,\|\vec{\zeta}-i\vec{\eta}\|^2\right)=4\langle\vec{\zeta}\mid\vec{\eta}\rangle.$$

Q12(3) Prove the Schwarz inequality in $\vec{\mathbb{V}}^N$:[1]

$$|\langle\vec{\zeta}\mid\vec{\eta}\rangle|\le\|\vec{\zeta}\|\,\|\vec{\eta}\|.$$

SQ12(3) The Schwarz inequality clearly holds if either $\vec{\zeta}$ or $\vec{\eta}$ are zero, or if $\vec{\zeta}=\vec{\eta}$. Indeed we have an equality in these cases. Generally, when neither are zero we can define a new vector $\vec{\xi}$ by

$$\vec{\xi}=\vec{\zeta}-c\,\vec{\eta},\qquad c=\frac{\langle\vec{\eta}\mid\vec{\zeta}\rangle}{\langle\vec{\eta}\mid\vec{\eta}\rangle}=\frac{\langle\vec{\eta}\mid\vec{\zeta}\rangle}{\|\vec{\eta}\|^2}.$$

$$\begin{aligned}\langle\vec{\xi}\mid\vec{\xi}\rangle&=\langle\vec{\zeta}\mid\vec{\zeta}\rangle+c^*c\langle\vec{\eta}\mid\vec{\eta}\rangle-c\langle\vec{\zeta}\mid\vec{\eta}\rangle-c^*\langle\vec{\eta}\mid\vec{\zeta}\rangle\\&=\|\vec{\zeta}\|^2+\frac{|\langle\vec{\eta}\mid\vec{\zeta}\rangle|^2}{\|\vec{\eta}\|^4}\|\vec{\eta}\|^2-\frac{\langle\vec{\eta}\mid\vec{\zeta}\rangle}{\|\vec{\eta}\|^2}\langle\vec{\zeta}\mid\vec{\eta}\rangle-\frac{\langle\vec{\eta}\mid\vec{\zeta}\rangle^*}{\|\vec{\eta}\|^2}\langle\vec{\eta}\mid\vec{\zeta}\rangle\\&=\|\vec{\zeta}\|^2+\frac{|\langle\vec{\eta}\mid\vec{\zeta}\rangle|^2}{\|\vec{\eta}\|^2}-\frac{|\langle\vec{\eta}\mid\vec{\zeta}\rangle|^2}{\|\vec{\eta}\|^2}-\frac{|\langle\vec{\eta}\mid\vec{\zeta}\rangle|^2}{\|\vec{\eta}\|^2}\\&=\|\vec{\zeta}\|^2-\frac{|\langle\vec{\eta}\mid\vec{\zeta}\rangle|^2}{\|\vec{\eta}\|^2}.\end{aligned}$$

The Schwarz inequality follows immediately, i.e.,

$$\begin{aligned}\langle\vec{\xi}\mid\vec{\xi}\rangle\ge0\;&\Rightarrow\;\|\vec{\zeta}\|^2-|\langle\vec{\zeta}\mid\vec{\eta}\rangle|^2/\|\vec{\eta}\|^2\ge0\\&\Rightarrow\;\|\vec{\zeta}\|\,\|\vec{\eta}\|\ge|\langle\vec{\zeta}\mid\vec{\eta}\rangle|.\end{aligned}$$

Q12(4) Prove the following triangle inequalities in $\vec{\mathbb{V}}^N$:[2]

$$\|\vec{\zeta}+\vec{\eta}\|\le\|\vec{\zeta}\|+\|\vec{\eta}\|.$$

$$\Big|\,\|\vec{\zeta}\|-\|\vec{\eta}\|\,\Big|\le\|\vec{\zeta}-\vec{\eta}\|.$$

SQ12(4) For the first triangle inequality we have

$$\begin{aligned}\|\vec{\zeta}+\vec{\eta}\|^2&=\langle\vec{\zeta}+\vec{\eta}\mid\vec{\zeta}+\vec{\eta}\rangle\\&=\langle\vec{\zeta}\mid\vec{\zeta}\rangle+\langle\vec{\eta}\mid\vec{\eta}\rangle+\langle\vec{\zeta}\mid\vec{\eta}\rangle+\langle\vec{\eta}\mid\vec{\zeta}\rangle\\&=\|\vec{\zeta}\|^2+\|\vec{\eta}\|^2+2\times\text{ real part of }\langle\vec{\eta}\mid\vec{\zeta}\rangle\\&\le\|\vec{\zeta}\|^2+\|\vec{\eta}\|^2+2\times|\langle\vec{\eta}\mid\vec{\zeta}\rangle|.\end{aligned}$$

---

[1] Roman p. 419. Prugovečki pp. 19–20.

[2] Work out $\|\vec{\zeta}+\vec{\eta}\|^2=\langle\vec{\zeta}+\vec{\eta}\mid\vec{\zeta}+\vec{\eta}\rangle$ and $\|\vec{\zeta}-\vec{\eta}\|^2=\langle\vec{\zeta}-\vec{\eta}\mid\vec{\zeta}-\vec{\eta}\rangle$. Note that $\langle\vec{\zeta}\mid\vec{\eta}\rangle+\langle\vec{\eta}\mid\vec{\zeta}\rangle=2\times$ the real part of $\langle\vec{\zeta}\mid\vec{\eta}\rangle$ which is less than or equal to $2|\langle\vec{\zeta}\mid\vec{\eta}\rangle|$.

The Schwarz inequality tells us that $|\langle \vec{\eta} \mid \vec{\zeta} \rangle| \le ||\vec{\zeta}\,||\,||\vec{\eta}\,||$. So

$$||\vec{\zeta} + \vec{\eta}\,||^2 \le ||\vec{\zeta}\,||^2 + ||\vec{\eta}\,||^2 + 2\,||\vec{\zeta}\,||\,||\vec{\eta}\,|| = \big(||\vec{\zeta}\,|| + ||\vec{\eta}\,||\big)^2,$$
$$\Rightarrow ||\vec{\zeta} + \vec{\eta}\,|| \le ||\vec{\zeta}\,|| + ||\vec{\eta}\,||.$$

For the second triangle inequality we have

$$\begin{aligned} ||\vec{\zeta} - \vec{\eta}\,||^2 &= \langle \vec{\zeta} - \vec{\eta} \mid \vec{\zeta} - \vec{\eta} \rangle \\ &= \langle \vec{\zeta} \mid \vec{\zeta} \rangle + \langle \vec{\eta} \mid \vec{\eta} \rangle - \langle \vec{\zeta} \mid \vec{\eta} \rangle - \langle \vec{\eta} \mid \vec{\zeta} \rangle \\ &= \langle \vec{\zeta} \mid \vec{\zeta} \rangle + \langle \vec{\eta} \mid \vec{\eta} \rangle - 2 \times \text{ real part of } \langle \vec{\eta} \mid \vec{\zeta} \rangle \\ &\ge ||\vec{\zeta}\,||^2 + ||\vec{\eta}\,||^2 - 2 \times |\langle \vec{\eta} \mid \vec{\zeta} \rangle|. \end{aligned}$$

Since $|\langle \vec{\eta} \mid \vec{\zeta} \rangle| \le ||\vec{\zeta}\,||\,||\vec{\eta}\,||$ we get

$$||\vec{\zeta} - \vec{\eta}\,||^2 \ge ||\vec{\zeta}\,||^2 + ||\vec{\eta}\,||^2 - 2||\vec{\zeta}\,||\,||\vec{\eta}\,|| = \big(||\vec{\zeta}|| - ||\vec{\eta}\,||\big)^2$$
$$\Rightarrow ||\vec{\zeta} - \vec{\eta}\,|| \ge \big|\,||\vec{\zeta}|| - ||\vec{\eta}\,||\,\big|.$$

Q12(5) Show that the Frobenius expression in Eq. (12.23) satisfies the properties CSP11.2.2(1), CSP11.2.2(2) and CSP11.2.2(3) of a scalar product.

SQ12(5) The Frobenius expression in Eq. (12.23) is

$$\langle \vec{\boldsymbol{M}} \mid \vec{\boldsymbol{N}} \rangle := M_{11}^* N_{11} + M_{12}^* N_{12} + M_{21}^* N_{21} + M_{22}^* N_{22}.$$

Property CSP11.2.2(1) is clearly satisfied. CSP11.2.2(2) is also satisfied since

$$\begin{aligned} \langle \vec{\boldsymbol{M}} \mid a\vec{\boldsymbol{N}} + b\vec{\boldsymbol{L}} \rangle &= M_{11}^*\big(aN_{11} + bL_{11}\big) + M_{12}^*\big(aN_{12} + bL_{12}\big) \\ &\quad + M_{21}^*\big(aN_{21} + bL_{21}\big) + M_{22}^*\big(aN_{22} + bL_{22}\big) \\ &= a\langle \vec{\boldsymbol{M}} \mid \vec{\boldsymbol{N}} \rangle + b\langle \vec{\boldsymbol{M}} \mid \vec{\boldsymbol{L}} \rangle. \end{aligned}$$

For property CPS11.2.2(3) we first observe that

$$\langle \vec{\boldsymbol{M}} \mid \vec{\boldsymbol{M}} \rangle = M_{11}^* M_{11} + M_{12}^* M_{12} + M_{21}^* M_{21} + M_{22}^* M_{22} \ge 0,$$

since each term is non-negative. Also we have

$$M_{11}^* M_{11} + M_{12}^* M_{12} + M_{21}^* M_{21} + M_{22}^* M_{22} = 0,$$

which implies each term must vanish, i.e., $\vec{\boldsymbol{M}} = \vec{\boldsymbol{0}}$.

Q12(6) Show that the integral expression in Eq. (12.25) satisfies the properties CSP11.2.2(1), CSP11.2.2(2) and CSP11.2.2(3) of a scalar product.

SQ12(6) The integral expression clearly satisfies CSP11.2.2(1). The linear property of integrals means that CSP11.2.2(2) is also satisfied, i.e., we have

$$\int_0^L \varphi^*(x)\big(a\phi(x) + b\psi(x)\big)\,dx$$
$$= a\int_0^L \varphi^*(x)\phi(x)\,dx + b\int_0^L \varphi^*(x)b\psi(x)\,dx.$$

For continuous functions we have

$$\int_0^L \phi^*(x)\phi(x)\,dx = 0 \;\Rightarrow\; \phi(x) = 0,$$

which is CSP11.2.2(3).

Q12(7) Verify that the vectors $\vec{\varphi}_\ell$ corresponding to the functions $\varphi_\ell(x)$ in Eq. (12.29) are orthonormal.

SQ12(7) For $\ell = \ell'$ the integrand becomes $1/L$. Integrating over 0 and L the integral gives the value 1. For $\ell \neq \ell'$ we have

$$\int_0^L \varphi_\ell^*(x)\varphi_{\ell'}(x)\,dx = \frac{1}{L}\int_0^L e^{2\pi i(\ell'-\ell)x/L}dx$$
$$= \frac{1}{L}\frac{L}{2\pi i(\ell'-\ell)}e^{2\pi i(\ell'-\ell)x/L}\Big|_0^L$$
$$= \frac{1}{2\pi i(\ell'-\ell)}(1-1) = 0,$$

bearing in mind that $\exp(2\pi ni) = 1$ for any positive or negative integer n. Hence their corresponding vectors $\vec{\varphi}_\ell$ are orthonormal.

Q12(8) Show that two orthogonal vectors are linearly independent.

SQ12(8) The proof follows that of SQ(6.4). Let $\vec{\zeta}_1$ and $\vec{\zeta}_2$ be two orthogonal vectors and c_1 and c_2 be two complex numbers such that $c_1\vec{\zeta}_1 + c_2\vec{\zeta}_2 = \vec{0}$. Taking scalar product of this equation with $\vec{\zeta}_1$ and $\vec{\zeta}_2$ in turn we get, on account of $\langle\vec{\zeta}_1 \mid \vec{\zeta}_2\rangle = \langle\vec{\zeta}_2 \mid \vec{\zeta}_1\rangle = 0$,

$$c_1\langle\vec{\zeta}_1 \mid \vec{\zeta}_1\rangle = 0 \;\Rightarrow\; c_1 = 0 \;\text{ and }\; c_2\langle\vec{\zeta}_2 \mid \vec{\zeta}_2\rangle = 0 \;\Rightarrow\; c_2 = 0.$$

The two vectors are therefore linearly independent.

Q12(9) Show that the three Pauli matrices σ_x, σ_y and σ_z in Eq. (7.9) together with the 2×2 identity matrix form a basis for the vector space of 2 × 2 complex matrices with Frobenious scalar product. Is this an orthonormal basis?

SQ12(9) The set of 2 × 2 matrices constitutes a four-dimensional scalar product space with Frobenius scalar product. One can verify that the Pauli matrices and the 2 × 2 identity matrix $\boldsymbol{I}_{2\times 2}$ form an orthogonal set with respect to the Frobenius scalar product, e.g.,

$$\langle \sigma_x \mid \sigma_y \rangle = 0 \times 0 + 1 \times (-i) + 1 \times i + 0 \times 0 = 0.$$

$$\langle \sigma_y \mid \sigma_z \rangle = 0 \times 1 + (-i)^* \times 0 + i^* \times 0 + 0 \times 1 = 0.$$

$$\langle \sigma_z \mid \sigma_x \rangle = 1 \times 0 + 0 \times 1 + 0 \times 1 + (-1) \times 0 = 0.$$

$$\langle \sigma_x \mid \boldsymbol{I}_{2\times 2} \rangle = 0 \times 1 + 1 \times 0 + 1 \times 0 + 0 \times 1 = 0.$$

$$\langle \sigma_y \mid \boldsymbol{I}_{2\times 2} \rangle = 0 \times 1 + (-i)^* \times 0 + i^* \times 0 + 0 \times 1 = 0.$$

$$\langle \sigma_z \mid \boldsymbol{I}_{2\times 2} \rangle = 1 \times 1 + 0 \times 0 + 0 \times 0 + (-1) \times 1 = 0.$$

It follows that the Pauli matrices and the 2 × 2 identity matrix $\boldsymbol{I}_{2\times 2}$ form a linearly independent set, i.e., they form a basis for the four-dimensional space of the set of 2 × 2 matrices. The basis is not orthonormal since Pauli matrices are not normalised in Frobenius scalar product.

Q12(10) Show that the mapping in Eq. (12.32) is unitary.

SQ12(10) The space $\vec{\mathbb{C}}^N$ is spanned by $\vec{e}^{\,c}_\ell$. It follows that f in Eq. (12.32) is a mapping of $\vec{\mathbb{V}}^N$ onto $\vec{\mathbb{C}}^N$. The mapping is also one-to-one since the coefficients c_ℓ associate every vector $\vec{\phi}$ in $\vec{\mathbb{V}}^N$ a unique vector $f(\vec{\phi}) \in \vec{\mathbb{C}}^N$. In other words f establishes an isomorphism between $\vec{\mathbb{V}}^N$ and $\vec{\mathbb{C}}^N$.

The mapping also preserves the scalar product since for any given $\vec{\phi} = \sum_\ell c_\ell \vec{\varphi}_\ell$ and $\vec{\phi}\,' = \sum_\ell c'_\ell \vec{\varphi}_\ell$ in $\vec{\mathbb{V}}^N$ we have

$$\langle \vec{\phi} \mid \vec{\phi}\,' \rangle = \sum_{\ell=1}^{N} c^*_\ell c'_\ell \quad \text{and}$$

$$\langle f(\vec{\phi}) \mid f(\vec{\phi}\,') \rangle = \langle \sum_\ell c_\ell \vec{e}^{\,c}_\ell \mid \sum_{\ell'} c'_\ell \vec{e}^{\,c}_{\ell'} \rangle = \sum_{\ell=1}^{N} c^*_\ell c'_\ell .$$

The mapping is therefore unitary.

Chapter 13

Operators on *N*-Dimensional Complex Vectors

Q13(1) Prove Eqs. (13.3) and (13.4).

SQ13(1) The proofs are similar to that given in SQ12(8). We have, for all $\vec{\zeta},\ \vec{\eta} \in \vec{\mathbb{V}}^N$,

$$\begin{aligned}\langle \vec{\zeta} \mid (c_1\widehat{A}_1 + c_2\widehat{A}_2)\vec{\eta}\,\rangle &= \langle \vec{\zeta} \mid c_1\widehat{A}_1\vec{\eta}\,\rangle + \langle \vec{\zeta} \mid c_2\widehat{A}_2\vec{\eta}\,\rangle \\ &= \langle c_1^*\widehat{A}_1^\dagger\vec{\zeta} \mid \vec{\eta}\,\rangle + \langle c_2^*\widehat{A}_2^\dagger\vec{\zeta} \mid \vec{\eta}\,\rangle \\ &= \langle (c_1^*\widehat{A}_1^\dagger + c_2^*\widehat{A}_2^\dagger)\vec{\zeta} \mid \vec{\eta}\,\rangle \\ \Rightarrow \quad (c_1\widehat{A}_1 + c_2\widehat{A}_2)^\dagger &= c_1^*\widehat{A}_1^\dagger + c_2^*\widehat{A}_2^\dagger.\end{aligned}$$

$$\begin{aligned}\langle \vec{\zeta} \mid (\widehat{A}\widehat{B})\vec{\eta}\,\rangle = \langle \vec{\zeta} \mid \widehat{A}(\widehat{B}\vec{\eta})\rangle &= \langle \widehat{A}^\dagger\vec{\zeta} \mid \widehat{B}\vec{\eta}\,\rangle = \langle \widehat{B}^\dagger\widehat{A}^\dagger\vec{\zeta} \mid \vec{\eta}\,\rangle \\ &= \langle (\widehat{B}^\dagger\widehat{A}^\dagger)\vec{\zeta} \mid \vec{\eta}\,\rangle \\ \Rightarrow \quad (\widehat{A}\widehat{B})^\dagger &= \widehat{B}^\dagger\widehat{A}^\dagger.\end{aligned}$$

Q13(2) Verify Eq. (13.7).

SQ13(2) Evaluating $\langle \vec{\zeta} + \vec{\eta} \mid \widehat{A}(\vec{\zeta} + \vec{\eta})\rangle$ and $\langle \vec{\zeta} + i\vec{\eta} \mid \widehat{A}(\vec{\zeta} + i\vec{\eta})\,\rangle$ we get

$$\langle \vec{\zeta} + \vec{\eta} \mid \widehat{A}(\vec{\zeta} + \vec{\eta})\rangle = \langle \vec{\zeta} \mid \widehat{A}\vec{\zeta}\,\rangle + \langle \vec{\eta} \mid \widehat{A}\vec{\eta}\,\rangle + \langle \vec{\zeta} \mid \widehat{A}\vec{\eta}\,\rangle + \langle \vec{\eta} \mid \widehat{A}\vec{\zeta}\,\rangle.$$

$$\langle \vec{\zeta} + i\vec{\eta} \mid \widehat{A}(\vec{\zeta} + i\vec{\eta})\,\rangle = \langle \vec{\zeta} \mid \widehat{A}\vec{\zeta}\,\rangle + \langle \vec{\eta} \mid \widehat{A}\vec{\eta}\,\rangle + i\langle \vec{\zeta} \mid \widehat{A}\vec{\eta}\,\rangle - i\langle \vec{\eta} \mid \widehat{A}\vec{\zeta}\,\rangle.$$

Quantum Mechanics: Problems and Solutions
K. Kong Wan

ISBN 978-981-4800-72-3 (Paperback), 978-0-429-29647-5 (eBook)
www.jennystanford.com

From the above equations we get

$$\langle \vec{\zeta} + \vec{\eta} \mid \widehat{A}(\vec{\zeta} + \vec{\eta})\rangle - i\langle \vec{\zeta} + i\vec{\eta} \mid \widehat{A}(\vec{\zeta} + i\vec{\eta})\rangle$$
$$= (1-i)\big(\langle \vec{\zeta} \mid \widehat{A}\vec{\zeta}\,\rangle + \langle \vec{\eta} \mid \widehat{A}\vec{\eta}\rangle\big) + 2\langle \vec{\zeta} \mid \widehat{A}\vec{\eta}\,\rangle.$$

Equation (13.7) follows immediately.

Q13(3) Consider the xy-plane as a two-dimensional real vector space in its own right. In basis $\{\vec{i},\ \vec{j}\,\}$ an arbitrary vector $\vec{v}$ on the xy-plane has the matrix representation

$$\boldsymbol{C}_{\vec{v}} = \begin{pmatrix} v_x \\ v_y \end{pmatrix}.$$

In basis $\{\vec{i},\ \vec{j}\,\}$ the operator $\widehat{R}_p(\pi/2)$ which rotates any vector $\vec{u}$ on the xy-plane about the origin by an angle of $\pi/2$ has the following matrix representation in accordance with Eq. (7.136):

$$\boldsymbol{R}_p(\pi/2) = \begin{pmatrix} 0 & -1 \\ 1 & 0 \end{pmatrix}.$$

Show that $\langle \vec{v} \mid \widehat{R}_p(\pi/2)\vec{v}\,\rangle = 0$ and explain why this result does not satisfy Theorem 13.1(1).

SQ13(3) We have, by Eqs. (7.136), (7.137) and (7.138),

$$\langle \vec{v} \mid \widehat{R}_p(\pi/2)\vec{v}\,\rangle = \langle \boldsymbol{C}_{\vec{v}} \mid \boldsymbol{R}_p(\pi/2)\boldsymbol{C}_{\vec{v}}\rangle = \boldsymbol{C}_{\vec{v}}^{\dagger} \cdot \boldsymbol{R}_p(\pi/2)\boldsymbol{C}_{\vec{v}}$$
$$= \begin{pmatrix} v_x^* & v_y^* \end{pmatrix} \begin{pmatrix} 0 & -1 \\ 1 & 0 \end{pmatrix} \begin{pmatrix} v_x \\ v_y \end{pmatrix}$$
$$= \begin{pmatrix} v_x & v_y \end{pmatrix} \begin{pmatrix} -v_y \\ v_x \end{pmatrix} = 0 \quad \text{since } v_x,\ v_y \text{ are real.}$$

This result does not satisfy Theorem 13.1(1), i.e., $\langle \vec{v} \mid \widehat{R}_p(\pi/2)\vec{v}\,\rangle = 0$ for all $\vec{v}$ in the x-y plane does not imply $\widehat{R}_p(\pi/2) = \hat{0}$. Theorem 13.1(1) is proved using Eq. (13.7) which applies to complex vector spaces. Hence the theorem does not apply to the x-y plane which is a real vector space unless the operator is selfadjoint (*see* Theorem 9.4.1(1)). As pointed out in C9.4.1(6) the operator $\widehat{R}_p(\pi/2)$ is not selfadjoint.

Q13(4) Prove that every vector $\vec{\zeta}$ in $\vec{\mathcal{V}}^N$ can be decomposed uniquely as a sum of a vector lying in a given subspace $\vec{\mathcal{S}}$ and another one lying in its orthogonal complement $\vec{\mathcal{S}}^{\perp}$ as shown in Eq. (13.15).

SQ13(4)

(1) For uniqueness let us suppose there are two such decompositions

$$\vec{\zeta} = \vec{\zeta}_{\vec{S}} + \vec{\zeta}_{\vec{S}^\perp} \quad \text{and} \quad \vec{\zeta} = \vec{\zeta}\,'_{\vec{S}} + \vec{\zeta}\,'_{\vec{S}^\perp}.$$

Then we have

$$\vec{0} = \left(\vec{\zeta}_{\vec{S}} - \vec{\zeta}\,'_{\vec{S}}\right) + \left(\vec{\zeta}_{\vec{S}^\perp} - \vec{\zeta}\,'_{\vec{S}^\perp}\right).$$

The two vectors $(\vec{\zeta}_{\vec{S}} - \vec{\zeta}\,'_{\vec{S}}) \in \vec{S}$ and $(\vec{\zeta}_{\vec{S}^\perp} - \vec{\zeta}\,'_{\vec{S}^\perp}) \in \vec{S}^\perp$ are orthogonal and hence linearly independent, i.e., their sum cannot be zero. It follows that the two vectors must be zero separately, i.e., we must have

$$\vec{\zeta}_{\vec{S}} = \vec{\zeta}\,'_{\vec{S}} \quad \text{and} \quad \vec{\zeta}_{\vec{S}^\perp} = \vec{\zeta}\,'_{\vec{S}^\perp}.$$

This proves the uniqueness of the decomposition.

(2) We need to show the existence of such a decomposition. Since $\vec{S}$ is a subspace there exists an orthonormal basis $\{\vec{\varepsilon}_j, j = 1, 2, \cdots, M \leq N\}$ for $\vec{S}$. Let

$$\vec{\zeta}_{\vec{S}} = \sum_{j=1}^{M} \langle \vec{\varepsilon}_j \mid \vec{\zeta}\, \rangle \vec{\varepsilon}_j \quad \text{and} \quad \vec{\zeta}_{\vec{S}^\perp} = \vec{\zeta} - \vec{\zeta}_{\vec{S}}.$$

Clearly $\vec{\zeta}_{\vec{S}}$ lies in the subspace $\vec{S}$. Also $\vec{\zeta}_{\vec{S}^\perp}$ lies in $\vec{S}^\perp$ since

$$\begin{aligned}\langle \vec{\varepsilon}_j \mid \vec{\zeta}_{\vec{S}^\perp} \rangle &= \langle \vec{\epsilon}_j \mid \vec{\zeta}\, \rangle - \langle \vec{\varepsilon}_j \mid \sum_{j'=1}^{M} \langle \vec{\varepsilon}_{j'} \mid \vec{\zeta}\, \rangle \vec{\varepsilon}_{j'} \rangle \\ &= \langle \vec{\varepsilon}_j \mid \vec{\zeta}\, \rangle - \sum_{j'=1}^{M} \langle \vec{\varepsilon}_{j'} \mid \vec{\zeta}\, \rangle \langle \vec{\varepsilon}_j \mid \vec{\varepsilon}_{j'} \rangle \\ &= \langle \vec{\varepsilon}_j \mid \vec{\zeta}\, \rangle - \sum_{j'=1}^{M} \langle \vec{\varepsilon}_{j'} \mid \vec{\zeta}\, \rangle \delta_{jj'} = 0.\end{aligned}$$

We have a desired decomposition, i.e., we have

$$\vec{\zeta}_{\vec{S}} \in \vec{S}, \quad \vec{\zeta}_{\vec{S}^\perp} \in \vec{S}^\perp \quad \text{and} \quad \vec{\zeta}_{\vec{S}} + \vec{\zeta}_{\vec{S}^\perp} = \vec{\zeta}.$$

Q13(5) Show that Eq. (13.20) is equivalent to Eq. (13.25) or Eq. (13.26) in defining the order relation of projectors.

SQ13(5)[1]

(1) Proof of Eq. (13.20) implying Eq. (13.25):
Let us start with the obvious result $\widehat{P}_{\vec{S}_1}\vec{\zeta} \in \vec{S}_1$ for all $\vec{\xi} \in \vec{\mathbb{V}}^N$. If $\vec{S}_1 \subset \vec{S}_2$ we have $\widehat{P}_{\vec{S}_1}\vec{\zeta} \in \vec{S}_2$ also. It follows that the projector $\widehat{P}_{\vec{S}_2}$ will have no effect on $\widehat{P}_{\vec{S}_1}\vec{\zeta}$, i.e.,

$$(\widehat{P}_{\vec{S}_2}\widehat{P}_{\vec{S}_1})\vec{\zeta} = \widehat{P}_{\vec{S}_2}(\widehat{P}_{\vec{S}_1}\vec{\zeta}\,) = \widehat{P}_{\vec{S}_1}\vec{\zeta} \quad \forall\vec{\zeta} \in \vec{\mathbb{V}}^N.$$

This means $\widehat{P}_{\vec{S}_2}\widehat{P}_{\vec{S}_1} = \widehat{P}_{\vec{S}_1}$. Secondly, if $\vec{S}_1 \subset \vec{S}_2$, projecting onto $\vec{S}_2$ and then projecting onto $\vec{S}_1$ is the same as projecting onto $\vec{S}_1$ alone. We can prove this by taking the adjoint of the equality $\widehat{P}_{\vec{S}_2}\widehat{P}_{\vec{S}_1} = \widehat{P}_{\vec{S}_1}$, i.e., we have

$$\widehat{P}_{\vec{S}_1} = \widehat{P}^{\dagger}_{\vec{S}_1} = (\widehat{P}_{\vec{S}_2}\widehat{P}_{\vec{S}_1})^{\dagger} = \widehat{P}^{\dagger}_{\vec{S}_1}\widehat{P}^{\dagger}_{\vec{S}_2} = \widehat{P}_{\vec{S}_1}\widehat{P}_{\vec{S}_2}.$$

The result shows that Eq. (13.20) implies Eq. (13.25).

(2) Proof of Eq. (13.25) implying Eq. (13.26):
Given $\widehat{P}_{\vec{S}_1} = \widehat{P}_{\vec{S}_1}\widehat{P}_{\vec{S}_2}$ we have $\widehat{P}_{\vec{S}_1}\vec{\eta} = \widehat{P}_{\vec{S}_1}\widehat{P}_{\vec{S}_2}\vec{\eta}$ for all $\vec{\eta} \in \vec{\mathbb{V}}^N$. Using Eq. (8.15) or Eq. (17.7) we immediately get[2]

$$||\widehat{P}_{\vec{S}_1}\vec{\eta}\,|| = ||\widehat{P}_{\vec{S}_1}(\widehat{P}_{\vec{S}_2}\vec{\eta})\,|| \le ||\widehat{P}_{\vec{S}_1}\,||\;||\widehat{P}_{\vec{S}_2}\vec{\eta}\,|| = ||\widehat{P}_{\vec{S}_2}\vec{\eta}\,||.$$

We have used the result that the norm of a projector is equal to 1, as stated in Theorem 13.2.2(1).

(3) Proof of Eq. (13.26) implying Eq. (13.20):
Let $\vec{\eta} = \vec{\eta}_{\vec{S}_1} + \vec{\eta}_{\vec{S}_1^{\perp}} = \widehat{P}_{\vec{S}_1}\vec{\eta} + \widehat{P}_{\vec{S}_1^{\perp}}\vec{\eta}$. Then we have

$$\langle\vec{\eta} \mid \vec{\eta}\rangle = ||\vec{\eta}\,||^2 = ||\widehat{P}_{\vec{S}_1}\vec{\eta}\,||^2 + ||\widehat{P}_{\vec{S}_1^{\perp}}\vec{\eta}\,||^2. \qquad (*)$$

Equation (∗) above implies:

(a) If $\vec{\eta} \in \vec{S}_1$, i.e., $\widehat{P}_{\vec{S}_1^{\perp}}\vec{\eta} = \vec{0}$, then $||\vec{\eta}\,|| = ||\widehat{P}_{\vec{S}_1}\vec{\eta}\,||$.
(b) If $||\vec{\eta}\,|| = ||\widehat{P}_{\vec{S}_1}\vec{\eta}\,||$, then $||\widehat{P}_{\vec{S}_1^{\perp}}\vec{\eta}\,|| = 0$ which implies $\widehat{P}_{\vec{S}_1^{\perp}}\vec{\eta} = \vec{0}$.

---

[1] Prugovečki p. 202. Roman Vol. 2 p. 538, p. 569.
[2] *See* P.13.1(1) on the definition of norm of an operator on $\vec{\mathbb{V}}^N$.

We can conclude that

$$\vec{\eta} \in \vec{\mathcal{S}}_1 \quad \text{if and only if} \quad ||\vec{\eta}\,|| = ||\widehat{P}_{\vec{\mathcal{S}}_1}\vec{\eta}\,||.$$

Now assume Eq. (13.26) holds. Then we have, for any $\vec{\eta} \in \vec{\mathcal{S}}_1$,

$$||\vec{\eta}\,|| = ||\widehat{P}_{\vec{\mathcal{S}}_1}\vec{\eta}\,|| \le ||\widehat{P}_{\vec{\mathcal{S}}_2}\vec{\eta}\,||.$$

$$||\widehat{P}_{\vec{\mathcal{S}}_2}\vec{\eta}\,|| \le ||\widehat{P}_{\vec{\mathcal{S}}_2}||\,||\vec{\eta}\,|| = ||\vec{\eta}\,||.$$

The second inequality arises from Eq. (8.15) or Eq. (17.7). We can also arrive at this result from the fact that the length of any projection $||\widehat{P}_{\vec{\mathcal{S}}_2}\vec{\eta}\,||$ must be less or equal to the length $||\vec{\eta}\,||$ of the original vector $\vec{\eta}$. It follows that

$$||\vec{\eta}\,|| \le ||\widehat{P}_{\vec{\mathcal{S}}_2}\vec{\eta}\,|| \le ||\vec{\eta}\,|| \;\Rightarrow\; ||\widehat{P}_{\vec{\mathcal{S}}_2}\vec{\eta}\,|| = ||\vec{\eta}\,||.$$

This in turn implies $\vec{\eta} \in \vec{\mathcal{S}}_2$, i.e., any $\vec{\eta} \in \vec{\mathcal{S}}_1$ must also be in $\vec{\mathcal{S}}_2$. In other words we have $\vec{\mathcal{S}}_1 \subset \vec{\mathcal{S}}_2$.

We have therefore showed that

$$\text{Eq. (13.20)} \;\Rightarrow\; \text{Eq. (13.25)} \;\Rightarrow\; \text{Eq. (13.26)} \;\Rightarrow\; \text{Eq. (13.20)}.$$

Q13(6) Prove Eqs. (13.22) and (13.23).

SQ13(6)

(1) Equation (13.22), i.e., $\big(|\vec{\zeta}\,\rangle\langle\vec{\eta}\,|\big)^\dagger = |\vec{\eta}\,\rangle\langle\vec{\zeta}\,|$, is proved by the following scalar product calculation for any $\vec{\xi},\ \vec{\epsilon} \in \vec{\mathbb{V}}^N$:

$$\langle\vec{\xi} \mid \big(|\vec{\zeta}\,\rangle\langle\vec{\eta}\,|\big)\vec{\epsilon}\,\rangle = \langle\vec{\xi} \mid \langle\vec{\eta} \mid \vec{\epsilon}\,\rangle\,\vec{\zeta}\,\rangle = \langle\vec{\eta} \mid \vec{\epsilon}\,\rangle\,\langle\vec{\xi} \mid \vec{\zeta}\,\rangle,$$

$$\langle\big(|\vec{\eta}\,\rangle\langle\vec{\zeta}\,|\big)\vec{\xi} \mid \vec{\epsilon}\,\rangle = \langle\langle\vec{\zeta} \mid \vec{\xi}\,\rangle\,\vec{\eta} \mid \vec{\epsilon}\,\rangle = \langle\vec{\zeta} \mid \vec{\xi}\,\rangle^*\langle\vec{\eta} \mid \vec{\epsilon}\,\rangle$$

$$= \langle\vec{\xi} \mid \vec{\zeta}\,\rangle\langle\vec{\eta} \mid \vec{\epsilon}\,\rangle$$

$$\Rightarrow\quad \langle\big(|\vec{\eta}\,\rangle\langle\vec{\zeta}\,|\big)\vec{\xi} \mid \vec{\epsilon}\,\rangle = \langle\vec{\xi} \mid \big(|\vec{\zeta}\,\rangle\langle\vec{\eta}\,|\big)\vec{\epsilon}\,\rangle.$$

(2) Equation (13.23), i.e.,

$$\big(|\vec{\zeta}_1\rangle\langle\vec{\eta}_1|\big)\big(|\vec{\zeta}_2\rangle\langle\vec{\eta}_2|\big) = \langle\vec{\eta}_1 \mid \vec{\zeta}_2\rangle\,|\vec{\zeta}_1\rangle\langle\vec{\eta}_2|,$$

is proved in two steps. First we have, for any $\vec{\xi} \in \vec{\mathbb{V}}^N$,

$$\big(|\vec{\zeta}_1\rangle\langle\vec{\eta}_1|\big)\big(|\vec{\zeta}_2\rangle\langle\vec{\eta}_2|\big)\vec{\xi} = \big(|\vec{\zeta}_1\rangle\langle\vec{\eta}_1|\big)\langle\vec{\eta}_2 \mid \vec{\xi}\,\rangle\,\vec{\zeta}_2$$

$$= \langle\vec{\eta}_2 \mid \vec{\xi}\,\rangle\big(|\vec{\zeta}_1\rangle\langle\vec{\eta}_1|\big)\vec{\zeta}_2$$

$$= \langle\vec{\eta}_2 \mid \vec{\xi}\,\rangle\langle\vec{\eta}_1 \mid \vec{\zeta}_2\rangle\,\vec{\zeta}_1.$$

Secondly we have

$$\Big(\langle \vec{\eta}_1 \mid \vec{\zeta}_2 \rangle \, |\vec{\zeta}_1 \rangle\langle \vec{\eta}_2 \,|\Big)\vec{\xi} = \langle \vec{\eta}_1 \mid \vec{\zeta}_2 \rangle \, \langle \vec{\eta}_2 \mid \vec{\xi} \,\rangle \, \vec{\zeta}_1.$$

The required result follows immediately.

Q13(7) Prove statement (3) of Theorem 13.2.2(1).

SQ13(7) Let $\widehat{P}_{\vec{S}_1}$ and $\widehat{P}_{\vec{S}_2}$ be two projectors onto the subspaces $\vec{S}_1$ and $\vec{S}_2$ respectively. Let $\vec{\zeta}$ be an arbitrary vector, and let $\widehat{P}_{\vec{S}_1}\vec{\zeta} = \vec{\zeta}_1$ and $\widehat{P}_{\vec{S}_2}\vec{\zeta} = \vec{\zeta}_2$. By Definition 9.3.2(2) two projectors are orthogonal if the subspaces onto which they project are orthogonal.

(1) If $\widehat{P}_{\vec{S}_1}$ and $\widehat{P}_{\vec{S}_2}$ are orthogonal then $\vec{\zeta}_2$ has no projection onto $\vec{\zeta}_1$ since $\langle \vec{\zeta}_1 \mid \vec{\zeta}_2 \rangle = 0$.

$$(\widehat{P}_{\vec{S}_1}\widehat{P}_{\vec{S}_2})\vec{\zeta} = \widehat{P}_{\vec{S}_1}(\widehat{P}_{\vec{S}_2}\vec{\zeta}\,) = \widehat{P}_{\vec{S}_1}\vec{\zeta}_2 = \vec{0}.$$

(2) If $\widehat{P}_{\vec{S}_1}\widehat{P}_{\vec{S}_2} = \widehat{0}$ then

$$\widehat{P}_{\vec{S}_1}\widehat{P}_{\vec{S}_2}\vec{\zeta} = \widehat{P}_{\vec{S}_1}\vec{\zeta}_2 = \vec{0} \;\Rightarrow\; \vec{\zeta}_2 \text{ has no projection onto } \vec{S}_1.$$

Since $\widehat{P}_{\vec{S}_1}\widehat{P}_{\vec{S}_2} = \widehat{0} \Rightarrow \widehat{P}_{\vec{S}_2}\widehat{P}_{\vec{S}_1} = \widehat{0}$ we can similarly deduce that $\vec{\zeta}_1$ has no projection onto $\vec{S}_2$. In other words $\vec{S}_1$ and $\vec{S}_2$ are orthogonal.

Q13(8) Prove the two expressions in Eq. (13.29).

SQ13(8) We have, for any vector $\vec{\eta}$,

$$\vec{\eta} = \sum_{\ell=1}^{N} c_\ell \, \vec{\varepsilon}_\ell, \quad c_\ell = \langle \vec{\varepsilon}_\ell \mid \vec{\eta} \rangle \;\Rightarrow\; \langle \vec{\eta} \mid \vec{\eta} \rangle = \sum_{\ell=1}^{N} |\langle \vec{\varepsilon}_\ell \mid \vec{\eta} \rangle|^2.$$

We also have

$$\sum_{\ell=1}^{N} \langle \vec{\eta} \mid \widehat{P}_{\vec{\varepsilon}_\ell} \vec{\eta} \rangle = \sum_{\ell=1}^{N} \langle \vec{\eta} \mid \langle \vec{\varepsilon}_\ell \mid \vec{\eta} \,\rangle \, \vec{\varepsilon}_\ell \rangle = \sum_{\ell=1}^{N} |\langle \vec{\varepsilon}_\ell \mid \vec{\eta} \rangle|^2.$$

It follows that

$$\langle \vec{\eta} \mid \left(\sum_{\ell=1}^{N} \widehat{P}_{\vec{\varepsilon}_\ell} \right) \vec{\eta} \rangle = \sum_{\ell=1}^{N} \langle \vec{\eta} \mid \widehat{P}_{\vec{\varepsilon}_\ell} \vec{\eta} \rangle = \langle \vec{\eta} \mid \vec{\eta} \rangle.$$

We can deduce two results:

(1) For any unit vector $\vec{\eta}$ we have

$$\langle \vec{\eta} \mid \left(\sum_{\ell=1}^{N} \widehat{P}_{\vec{\varepsilon}_\ell} \right) \vec{\eta} \rangle = 1.$$

(2) We also have

$$\sum_{\ell=1}^{N} \widehat{P}_{\vec{\varepsilon}_\ell} = \widehat{I\!I} \quad \text{on account of Eq. (13.10).}$$

This is a major result stated in Eq. (13.28).

Next, for any unit vector $\vec{\zeta}$ we have by Schwarz inequality[3]

$$|\langle \vec{\zeta} \mid \widehat{P}_{\vec{\varepsilon}_\ell} \vec{\zeta} \rangle| \leq || \vec{\zeta} || \, || \widehat{P}_{\vec{\varepsilon}_\ell} \vec{\zeta} || = || \widehat{P}_{\vec{\varepsilon}_\ell} \vec{\zeta} || \leq || \widehat{P}_{\vec{\varepsilon}_\ell} || = 1.$$

We also have

$$\langle \vec{\zeta} \mid \widehat{P}_{\vec{\varepsilon}_\ell} \vec{\zeta} \rangle = \langle \vec{\zeta} \mid \widehat{P}^2_{\vec{\varepsilon}_\ell} \vec{\zeta} \rangle = |\langle \widehat{P}_{\vec{\varepsilon}_\ell} \vec{\zeta} \mid \widehat{P}_{\vec{\varepsilon}_\ell} \vec{\zeta} \rangle = || \widehat{P}_{\vec{\varepsilon}_\ell} \vec{\zeta} ||^2 \geq 0.$$

The required result follows, i.e.,

$$\langle \vec{\zeta} \mid \widehat{P}_{\vec{\varepsilon}_\ell} \vec{\zeta} \rangle \in [0, 1].$$

Q13(9) Prove Theorem 13.3.1(1) on selfadjoint operators.

SQ13(9) If $\widehat{A}$ is selfadjoint, then, using the properties of scalar product, we get

$$\langle \vec{\zeta} \mid \widehat{A} \vec{\zeta} \rangle^* = \langle \widehat{A} \vec{\zeta} \mid \vec{\zeta} \rangle = \langle \vec{\zeta} \mid \widehat{A} \vec{\zeta} \rangle \quad \Rightarrow \quad \langle \vec{\zeta} \mid \widehat{A} \vec{\zeta} \rangle \in I\!R.$$

For the converse we first apply Eq. (13.7) we get

$$\langle \vec{\zeta} \mid \widehat{A} \vec{\eta} \rangle = \frac{1}{2} \Big\{ \langle \vec{\zeta} + \vec{\eta} \mid \widehat{A}(\vec{\zeta} + \vec{\eta}) \rangle - i \langle \vec{\zeta} + i\vec{\eta} \mid \widehat{A}(\vec{\zeta} + i\vec{\eta}) \rangle$$
$$+ (i-1)\big(\langle \vec{\zeta} \mid \widehat{A} \vec{\zeta} \rangle + \langle \vec{\eta} \mid \widehat{A} \vec{\eta} \rangle \big) \Big\},$$

$$\langle \vec{\eta} \mid \widehat{A} \vec{\zeta} \rangle = \frac{1}{2} \Big\{ \langle \vec{\eta} + \vec{\zeta} \mid \widehat{A}(\vec{\eta} + \vec{\zeta}) \rangle - i \langle \vec{\eta} + i\vec{\zeta} \mid \widehat{A}(\vec{\eta} + i\vec{\zeta}) \rangle$$
$$+ (i-1)\big(\langle \vec{\eta} \mid \widehat{A} \vec{\eta} \rangle + \langle \vec{\zeta} \mid \widehat{A} \vec{\zeta} \rangle \big) \Big\}.$$

Now suppose $\langle \vec{\xi} \mid \widehat{A} \vec{\xi} \rangle \in I\!R$ for any $\vec{\xi}$, e.g.,

$$\langle \vec{\zeta} + i\vec{\eta} \mid \widehat{A}(\vec{\zeta} + i\vec{\eta}) \rangle \in I\!R, \quad \langle \vec{\eta} + i\vec{\zeta} \mid \widehat{A}(\vec{\eta} + i\vec{\zeta}) \rangle \in I\!R.$$

---

[3] *See* Q12(3) and its solutions. We also use Eq. (8.15) or Eq. (17.7) to get $|| \widehat{P}_{\vec{\varepsilon}_\ell} \vec{\zeta} || \leq || \widehat{P}_{\vec{\varepsilon}_\ell} || \, || \vec{\zeta} || = || \widehat{P}_{\vec{\varepsilon}_\ell} ||$.

Then

$$\begin{aligned}\langle \hat{A}\vec{\zeta} \mid \vec{\eta}\,\rangle &= \langle \vec{\eta} \mid \hat{A}\vec{\zeta}\,\rangle^* \\ &= \frac{1}{2}\Big\{\langle \vec{\eta}+\vec{\zeta} \mid \hat{A}(\vec{\eta}+\vec{\zeta}\,)\rangle + i\langle \vec{\eta}+i\vec{\zeta} \mid \hat{A}(\vec{\eta}+i\vec{\zeta}\,)\rangle \\ &\quad -(i+1)\big(\langle \vec{\eta} \mid \hat{A}\vec{\eta}\,\rangle + \langle \vec{\zeta} \mid \hat{A}\vec{\zeta}\,\rangle\big)\Big\}.\end{aligned}$$

By expanding the scalar product expressions on the right-hand sides of the equations for $\langle \vec{\zeta} \mid \hat{A}\vec{\eta}\rangle$ and $\langle \hat{A}\vec{\zeta} \mid \vec{\eta}\,\rangle$ we get, for all $\vec{\zeta}, \vec{\eta}$,

$$\langle \hat{A}\vec{\zeta} \mid \vec{\eta}\,\rangle = \langle \vec{\zeta} \mid \hat{A}\vec{\eta}\rangle \;\Rightarrow\; \hat{A}^\dagger = \hat{A}.$$

Q13(10) Prove Spectral Theorems 13.3.2(1) and 13.3.2(2).

SQ13(10) We can prove Spectral Theorem 13.3.2(1) in the same way Theorem 9.4.5(1) is proved in the book. We know that $\hat{A}$ possesses a complete orthonormal set of eigenvectors $\vec{\varepsilon}_\ell$ and a corresponding complete orthogonal family of eigenprojectors $\hat{P}_{\vec{\varepsilon}_\ell}$ corresponding to a set of eigenvalues a_ℓ. Then for any vector $\vec{\zeta} \in \vec{\mathbb{V}}^N$ we have $\vec{\zeta} = \sum_{\ell=1}^3 \zeta_\ell \vec{\varepsilon}_\ell, \quad \zeta_\ell = \langle \vec{\varepsilon}_\ell \mid \vec{\zeta}\,\rangle$ and

$$\hat{A}\vec{\zeta} = \sum_{\ell=1}^{3} \zeta_\ell \hat{A}\vec{\varepsilon}_\ell = \sum_{\ell=1}^{3} a_\ell \zeta_\ell \vec{\varepsilon}_\ell.$$

Construct the operator

$$\hat{A}' := \sum_{\ell=1}^{3} a_\ell\, \hat{P}_{\vec{\varepsilon}_\ell}.$$

Then

$$\hat{A}'\vec{\zeta} = \sum_{\ell=1}^{3} a_\ell\, \hat{P}_{\vec{\varepsilon}_\ell}\vec{\zeta} = \sum_{\ell=1}^{3} a_\ell\, \langle \vec{\varepsilon}_\ell \mid \vec{v}\,\rangle \vec{\varepsilon}_\ell = \sum_{\ell=1}^{3} a_\ell \zeta_\ell \vec{\varepsilon}_\ell.$$

We see that $\hat{A}\vec{\zeta} = \hat{A}'\vec{\zeta}$ for every vector $\vec{\zeta}$. It follows that $\hat{A}' = \hat{A}$, i.e., the operator $\hat{A}$ has a spectral decomposition.

The spectral decomposition of $\hat{A}$ in Theorem 13.3.2(2) is just a restatement of that of Theorem 13.3.2(1) by grouping all the projectors of degenerate eigenvalues.

Q13(11) Show that the eigenprojectors of a selfadjoint operator commutes with each other and that a selfadjoint operator commutes with its eigenprojectors.

SQ13(11) The eigenprojectors $\widehat{P}_{\vec{\varepsilon}_\ell}$ of a selfadjoint operator $\widehat{A}$ in Theorem 13.3.2(1) are mutually orthogonal, i.e., $\widehat{P}_{\vec{\varepsilon}_\ell}\widehat{P}_{\vec{\varepsilon}_{\ell'}} = \widehat{0} = \widehat{P}_{\vec{\varepsilon}_{\ell'}}\widehat{P}_{\vec{\varepsilon}_\ell}$ for $\ell \neq \ell'$, and hence they mutually commute, i.e., $[\widehat{P}_{\vec{\varepsilon}_\ell}, \widehat{P}_{\vec{\varepsilon}_{\ell'}}] = \widehat{0}$. Substituting the spectral decomposition

$$\widehat{A} = \sum_{\ell=1}^{3} a_\ell \, \widehat{P}_{\vec{\varepsilon}_\ell}.$$

into the commutator $[\widehat{A}, \widehat{P}_{\vec{\varepsilon}_{\ell'}}]$ we get

$$[\widehat{A}, \widehat{P}_{\vec{\varepsilon}_{\ell'}}] = \sum_{\ell=1}^{3} a_\ell \, [\widehat{P}_{\vec{\varepsilon}_\ell}, \widehat{P}_{\vec{\varepsilon}_{\ell'}}] = \widehat{0}.$$

The same proof applies to the eigenprojectors $\widehat{P}^{\hat{A}}(a_m)$ of a selfadjoint operator $\widehat{A}$ in Theorem 13.3.2(2).

Q13(12) For a selfadjoint operator $\widehat{A}$ with its spectral decomposition given by Eq. (13.36) show that

$$\widehat{A}\widehat{P}^{\hat{A}}(a_m) = a_m \widehat{P}^{\hat{A}}(a_m).$$

SQ13(12) From the orthogonal nature of eigenprojectors we get

$$\begin{aligned}
\widehat{A}\,\widehat{P}^{\hat{A}}(a_m) &= \Big(\sum_{m'} a_{m'}\,\widehat{P}^{\hat{A}}(a_{m'})\Big)\,\widehat{P}^{\hat{A}}(a_m) \\
&= \sum_{m'} a_{m'}\,\widehat{P}^{\hat{A}}(a_{m'})\widehat{P}^{\hat{A}}(a_m) \\
&= \sum_{m'} a_{m'}\,\widehat{P}^{\hat{A}}(a_{m'})\delta_{m'm} = a_m \widehat{P}^{\hat{A}}(a_m),
\end{aligned}$$

which is the desired result.

Q13(13) Prove Eq. (13.50).

SQ13(13) To prove Eq. (13.50) for the expression for the adjoint we have, for any $\vec{\zeta}$ and $\vec{\eta}$

$$\begin{aligned}
\langle \vec{\xi} \mid e^{i\hat{A}}\,\vec{\eta}\,\rangle &= \langle \vec{\xi} \mid \sum_m e^{ia_m}\,\widehat{P}^{\hat{A}}(a_m)\,\vec{\eta}\,\rangle = \sum_m \langle \vec{\xi} \mid e^{ia_m}\,\widehat{P}^{\hat{A}}(a_m)\,\vec{\eta}\,\rangle \\
&= \sum_m \langle e^{-ia_m}\,\widehat{P}^{\hat{A}}(a_m)\,\vec{\xi} \mid \vec{\eta}\,\rangle = \langle \sum_m e^{-ia_m}\,\widehat{P}^{\hat{A}}(a_m)\,\vec{\xi} \mid \vec{\eta}\,\rangle \\
&= \langle e^{-i\hat{A}}\,\vec{\xi} \mid \vec{\eta}\,\rangle \;\Rightarrow\; \big(e^{i\hat{A}}\big)^\dagger = e^{-i\hat{A}}.
\end{aligned}$$

We have used Eqs. (13.48) and (13.49).

For the inverse we have, again using Eqs. (13.48) and (13.49),

$$\Big(\sum_m e^{ia_m}\,\widehat{P}^{\hat{A}}(a_m)\Big)\Big(\sum_{m'} e^{-ia_{m'}}\,\widehat{P}^{\hat{A}}(a_{m'})\Big)$$
$$=\sum_{m,m'} e^{ia_m}e^{-ia_{m'}}\,\widehat{P}^{\hat{A}}(a_{m'})\widehat{P}^{\hat{A}}(a_m)$$
$$=\sum_{m,m'} e^{ia_m}e^{-ia_{m'}}\,\widehat{P}^{\hat{A}}(a_{m'})\delta_{m'm}$$
$$=\sum_m \widehat{P}^{\hat{A}}(a_m)=\widehat{I\!I}$$
$$\Rightarrow \Big(\sum_m e^{ia_m}\,\widehat{P}^{\hat{A}}(a_m)\Big)^{-1}=\sum_{m'} e^{-ia_{m'}}\,\widehat{P}^{\hat{A}}(a_{m'})$$
$$\Rightarrow \big(e^{i\hat{A}}\big)^{-1}=e^{-i\hat{A}} \;\Rightarrow\; \big(e^{i\hat{A}}\big)^{\dagger}=\big(e^{i\hat{A}}\big)^{-1}=e^{-i\hat{A}}.$$

Q13(14) Prove Theorem 13.3.4(1).

SQ13(14) Theorem 13.3.4(1) can be proved in two steps.

(1) Let the spectral decompositions of two selfadjoint operators $\widehat{A}$ and $\widehat{B}$ be $\widehat{A}=\sum_m a_m\widehat{P}^{\hat{A}}(a_m)$ and $\widehat{B}=\sum_{m'} b_{m'}\widehat{P}^{\hat{B}}(b_{m'})$. Then

$$\widehat{A}\widehat{B}=\sum_{m,m'} a_m b_{m'}\,\widehat{P}^{\hat{A}}(a_m)\widehat{P}^{\hat{B}}(b_{m'}),$$
$$\widehat{B}\widehat{A}=\sum_{m,m'} a_m b_{m'}\,\widehat{P}^{\hat{B}}(b_{m'})\widehat{P}^{\hat{A}}(a_m),$$

and

$$[\widehat{A},\,\widehat{B}\,]=\sum_{m,m'} a_m b_{m'}\,\widehat{P}^{\hat{A}}(a_m)\widehat{P}^{\hat{B}}(b_{m'})-\sum_{m,m'} a_m b_{m'}\,\widehat{P}^{\hat{B}}(b_{m'})\widehat{P}^{\hat{A}}(a_m)$$
$$=\sum_{m,m'} a_m b_{m'}\,[\widehat{P}^{\hat{A}}(a_m),\,\widehat{P}^{\hat{B}}(b_{m'})].$$

It follows that

$$[\widehat{P}^{\hat{A}}(a_m),\,\widehat{P}^{\hat{B}}(b_{m'})]=\widehat{0}\;\forall\, m,\,m' \;\Rightarrow\; [\widehat{A},\,\widehat{B}\,]=\widehat{0}.$$

(2) To prove the converse let us assume that $[\widehat{A},\,\widehat{B}\,]=\widehat{0}$. From the result $\widehat{A}\widehat{P}^{\hat{A}}(a_m)=a_m\widehat{P}^{\hat{A}}(a_m)$ in Eq. (13.118) proved in SQ13(12) we deduce that

$$\widehat{A}\big(\widehat{B}\widehat{P}^{\hat{A}}(a_m)\vec{\zeta}\,\big)=\widehat{B}\big(\widehat{A}\widehat{P}^{\hat{A}}(a_m)\vec{\zeta}\,\big)=a_m\big(\widehat{B}\widehat{P}^{\hat{A}}(a_m)\vec{\zeta}\,\big).$$

This means that $\widehat{B}\widehat{P}^{\hat{A}}(a_m)\vec{\zeta}$ belongs to the eigensubspace of $\widehat{A}$ corresponding to the eigenvalue a_m. In other words we have

$$\widehat{P}^{\hat{A}}(a_m)\left(\widehat{B}\widehat{P}^{\hat{A}}(a_m)\vec{\zeta}\right) = \left(\widehat{B}\widehat{P}^{\hat{A}}(a_m)\vec{\zeta}\right) \ \forall\vec{\zeta}$$

$$\Rightarrow \ \left(\widehat{P}^{\hat{A}}(a_m)\widehat{B}\widehat{P}^{\hat{A}}(a_m)\right)\vec{\zeta} = \left(\widehat{B}\widehat{P}^{\hat{A}}(a_m)\right)\vec{\zeta} \ \forall\vec{\zeta}$$

$$\Rightarrow \ \widehat{P}^{\hat{A}}(a_m)\widehat{B}\widehat{P}^{\hat{A}}(a_m) = \widehat{B}\widehat{P}^{\hat{A}}(a_m).$$

Taking the adjoint of each side of the equation we get

$$\left(\widehat{P}^{\hat{A}}(a_m)\widehat{B}\widehat{P}^{\hat{A}}(a_m)\right)^{\dagger} = \left(\widehat{B}\widehat{P}^{\hat{A}}(a_m)\right)^{\dagger}.$$

$$\Rightarrow \ \widehat{P}^{\hat{A}}(a_m)\widehat{B}\widehat{P}^{\hat{A}}(a_m) = \widehat{P}^{\hat{A}}(a_m)\widehat{B}.$$

We get $\widehat{B}\widehat{P}^{\hat{A}}(a_m) = \widehat{P}^{\hat{A}}(a_m)\widehat{B}$, i.e.,

$$[\widehat{B},\, \widehat{P}^{\hat{A}}(a_m)] = \widehat{0} \ \text{ and similarly } \ [\widehat{A},\, \widehat{P}^{\hat{B}}(b_m)] = \widehat{0}.$$

What we have shown is that if a selfadjoint operator $\widehat{A}$ commutes with another selfadjoint operator $\widehat{B}$ then

(1) $\widehat{A}$ commutes with every eigenprojector $\widehat{P}^{\hat{B}}(b_m)$ of $\widehat{B}$, and
(2) $\widehat{B}$ commutes with every eigenprojector $\widehat{P}^{\hat{A}}(a_m)$ of $\widehat{A}$.

It follows that $\widehat{P}^{\hat{A}}(a_m)$ commutes with $\widehat{P}^{\hat{B}}(b_{m'})$ since $\widehat{P}^{\hat{A}}(a_m)$ commutes with $\widehat{B}$, i.e., $\widehat{P}^{\hat{A}}(a_m)$ will commute with all the eigenprojectors of $\widehat{B}$. Similarly the eigenprojectors $\widehat{P}^{\hat{B}}(b_m)$ of $\widehat{B}$ will commute with all the eigenprojectors $\widehat{P}^{\hat{A}}(a_{m'})$ of $\widehat{A}$.

Q13(15) Prove Eqs. (13.53) and (13.54).

SQ13(15)

(1) To prove Eq. (13.53) we start with

$$\sum_m \widehat{P}^{\hat{A}}(a_m) = \widehat{I\!I} \quad \text{and} \quad \sum_{m'} \widehat{P}^{\hat{B}}(b_{m'}) = \widehat{I\!I}.$$

Multiplying the above equations we immediately get

$$\Rightarrow \ \sum_{m,m'} \widehat{P}^{\hat{A}}(a_m)\widehat{P}^{\hat{B}}(b_{m'}) = \widehat{I\!I}.$$

(2) To prove Eq. (13.54) we can use Eq. (13.118) proved in SQ13(12), i.e., $\widehat{A}\widehat{P}^{\hat{A}}(a_m) = a_m\widehat{P}^{\hat{A}}(a_m)$, we get

$$\widehat{A}\vec{\eta}_{mm'} = \widehat{A}\widehat{P}^{\hat{A}}(a_m)\widehat{P}^{\hat{B}}(b_{m'})\vec{\eta} = a_m\widehat{P}^{\hat{A}}(a_m)\widehat{P}^{\hat{B}}(b_{m'})\vec{\eta}$$
$$= a_m\vec{\eta}_{mm'}.$$

A similar analysis shows that

$$\widehat{B}\vec{\eta}_{mm'} = b_{m'}\vec{\eta}_{mm'}.$$

Q13(16) Prove properties P13.4.1(1) to P13.4.1(4) of a unitary operator listed right after Theorem 13.4.1(1).

SQ13(16)

(1) P13.4.1(1) is true on account of Theorem 13.4.1(1). Explicitly we have, using Eq. (13.64),

$$\langle \widehat{U}\vec{\zeta} \mid \widehat{U}\vec{\eta}\,\rangle = \langle \widehat{U}^{\dagger}\widehat{U}\vec{\zeta} \mid \vec{\eta}\,\rangle = \langle \vec{\zeta} \mid \vec{\eta}\,\rangle.$$

(2) To prove P13.4.1(2) we have, for every vector $\vec{\zeta}$,

$$\langle \widehat{U}^{\dagger}\vec{\zeta} \mid \widehat{U}^{\dagger}\vec{\zeta}\,\rangle = \langle \vec{\zeta} \mid \widehat{U}\widehat{U}^{\dagger}\vec{\zeta}\,\rangle = \langle \vec{\zeta} \mid \vec{\zeta}\,\rangle.$$

Hence $\widehat{U}^{\dagger}$ is unitary. We have used Eq. (13.64) in the above calculation.

(3) To prove P13.4.1(3) we consider the eigenvalue equation of a unitary operator $\widehat{U}$, i.e., $\widehat{U}\vec{\eta} = \lambda\vec{\eta}$, where $\vec{\eta}$ is an eigenvector and λ is the associated eigenvalue. Then

$$\langle \widehat{U}\vec{\eta} \mid \widehat{U}\vec{\eta}\,\rangle = \langle \lambda\vec{\eta} \mid \lambda\vec{\eta}\,\rangle = \lambda^*\lambda\langle \vec{\eta} \mid \vec{\eta}\,\rangle.$$

Since a unitary operator preserves the scalar product we have

$$\langle \widehat{U}\vec{\eta} \mid \widehat{U}\vec{\eta}\,\rangle = \langle \vec{\eta} \mid \vec{\eta}\,\rangle.$$

It follows that $\lambda^*\lambda = |\lambda|^2 = 1$ which implies that λ is of the form $e^{ia}, a \in I\!R$.

(4) To prove P13.4.1(4) we start with the matrix elements $M_{\widehat{U}mn}$ of the matrix representation $\boldsymbol{M}_{\widehat{U}}$ of a given unitary operator $\widehat{U}$ in a given orthonormal basis $\{\vec{\varepsilon}_\ell\}$, i.e.,

$$M_{\widehat{U}mn} = \langle \vec{\varepsilon}_m \mid \widehat{U}\vec{\varepsilon}_n\rangle.$$

By definition the adjoint matrix $\boldsymbol{M}^{\dagger}_{\widehat{U}}$ has elements

$$M^{\dagger}_{\widehat{U}n\ell} = M^{*}_{\widehat{U}\ell n} = \langle \vec{\varepsilon}_\ell \mid \widehat{U}\,\vec{\varepsilon}_n \rangle^* = \langle \widehat{U}\,\vec{\varepsilon}_n \mid \vec{\varepsilon}_\ell \rangle.$$

The product matrix of $\boldsymbol{M}_{\widehat{U}}$ and $\boldsymbol{M}^{\dagger}_{\widehat{U}}$ has elements

$$\begin{aligned}\sum_n M_{\widehat{U}_{mn}} M^{\dagger}_{\widehat{U}_{n\ell}} &= \sum_n \langle \vec{\varepsilon}_m \mid \widehat{U}\,\vec{\varepsilon}_n \rangle \langle \widehat{U}\,\vec{\varepsilon}_n \mid \vec{\varepsilon}_\ell \rangle \\ &= \sum_n \langle \widehat{U}^{\dagger}\vec{\varepsilon}_m \mid \vec{\varepsilon}_n \rangle \langle \vec{\varepsilon}_n \mid \widehat{U}^{\dagger}\vec{\varepsilon}_\ell \rangle \\ &= \langle \widehat{U}^{\dagger}\vec{\varepsilon}_m \mid \widehat{U}^{\dagger}\vec{\varepsilon}_\ell \rangle = \langle \vec{\varepsilon}_m \mid \widehat{U}\,\widehat{U}^{\dagger}\vec{\varepsilon}_\ell \rangle = \delta_{m\ell} \\ &\Rightarrow \ \boldsymbol{M}_{\widehat{U}} \boldsymbol{M}^{\dagger}_{\widehat{U}} = \boldsymbol{I}.\end{aligned}$$

Hence matrix is unitary. We have used Eqs. (13.35) and (13.64).

Q13(17) Show that the operator in Eq. (13.69) is unitary.

SQ13(17) For the operator in Eq. (13.69) we have, using Eqs. (13.48), (13.50) and the orthogonal property of the eigenprojectors,

$$\widehat{U}^{\dagger} = \sum_{\ell=1}^{N} e^{-ia_\ell}\, \widehat{P}_{\vec{\varepsilon}_\ell}$$

$$\Rightarrow \ \widehat{U}^{\dagger}\widehat{U} = \sum_{\ell,m=1}^{N} e^{-ia_\ell} e^{ia_m}\, \widehat{P}_{\vec{\varepsilon}_\ell} \widehat{P}_{\vec{\varepsilon}_m} = \sum_{\ell}^{N} \widehat{P}_{\vec{\varepsilon}_\ell} = \widehat{I\!I}.$$

Hence $\widehat{U}$ is invertible (see Theorem 8.2.2(2) and P13.1(3)) with its inverse equal to its adjoint, i.e., $\widehat{U}^{\dagger} = \widehat{U}^{-1}$. It follows that $\widehat{U}$ is unitary by Theorem 13.4.1(1).

Q13(18) Show that the operator $\widehat{U}$ in Eq. (13.70) is unitary.

SQ13(18) For the operator in Eq. (13.70) we can write down its adjoint, using Eq. (13.22), i.e.,

$$\widehat{U}^{\dagger} = \sum_{\ell=1}^{N} \left| \vec{\varepsilon}_\ell \right\rangle \left\langle \vec{\varepsilon}\,'_\ell \right|.$$

Then

$$\begin{aligned}
\widehat{U}^\dagger \widehat{U} &= \Big(\sum_{\ell=1}^{N} |\vec{\varepsilon}_\ell\rangle\langle\vec{\varepsilon}\,'_\ell|\Big)\Big(\sum_{k=1}^{N} |\vec{\varepsilon}\,'_k\rangle\langle\vec{\varepsilon}_k|\Big) \\
&= \sum_{\ell,k=1}^{N} \Big(|\vec{\varepsilon}_\ell\rangle\langle\vec{\varepsilon}\,'_\ell|\Big)\Big(|\vec{\varepsilon}\,'_k\rangle\langle\vec{\varepsilon}_k|\Big) = \sum_{\ell,k=1}^{N} \langle\vec{\varepsilon}\,'_\ell \mid \vec{\varepsilon}\,'_k\rangle \Big(|\vec{\varepsilon}_\ell\rangle\rangle\langle\vec{\varepsilon}_k|\Big) \\
&= \sum_{k=1}^{N} |\vec{\varepsilon}_k\rangle\langle\vec{\varepsilon}_k| \quad \text{since } \langle\vec{\varepsilon}\,'_\ell \mid \vec{\varepsilon}\,'_k\rangle = \delta_{k\ell} \\
&= \widehat{I\!I}.
\end{aligned}$$

Hence $\widehat{U}$ is invertible (see SQ13(17)) with its adjoint equal to its inverse, i.e., $\widehat{U}^\dagger = \widehat{U}^{-1}$. Hence $\widehat{U}$ is unitary. We have used Eq. (13.23).

Q13(19) Prove Eq. (13.74) on the preservation of commutation relations.

SQ13(19) To prove the preservation of commutation relations in Eq. (13.75) we let $\widehat{A}' = \widehat{U}\,\widehat{A}\widehat{U}^\dagger$, $\widehat{B}' = \widehat{U}\,\widehat{B}\widehat{U}^\dagger$ and $\widehat{C}' = \widehat{U}\widehat{C}\widehat{U}^\dagger$. Then

$$\begin{aligned}
[\,\widehat{A}', \widehat{B}'] &= \widehat{A}'\widehat{B}' - \widehat{B}'\widehat{A}' = (\widehat{U}\,\widehat{A}\widehat{U}^\dagger)(\widehat{U}\,\widehat{B}\widehat{U}^\dagger) - (\widehat{U}\,\widehat{B}\widehat{U}^\dagger)(\widehat{U}\,\widehat{A}\widehat{U}^\dagger) \\
&= \widehat{U}\,\widehat{A}\widehat{B}\widehat{U}^\dagger - \widehat{U}\,\widehat{B}\widehat{A}\widehat{U}^\dagger = \widehat{U}\,(\widehat{A}\widehat{B} - \widehat{B}\widehat{A})\widehat{U}^\dagger = \widehat{U}\widehat{C}\widehat{U}^\dagger \\
&= \widehat{C}'.
\end{aligned}$$

Q13(20) Prove Eq. (13.77) on the preservation of the quadratic form.

SQ13(20) To prove the preservation of quadratic form in Eq. (13.75) we let $\widehat{A}' = \widehat{U}\,\widehat{A}\widehat{U}^\dagger$ and $\vec{\zeta}\,' = \widehat{U}\,\vec{\zeta}$. Then

$$\langle\vec{\zeta}\,' \mid \widehat{A}'\vec{\zeta}\,'\rangle = \langle\widehat{U}\vec{\zeta} \mid (\widehat{U}\,\widehat{A}\widehat{U}^\dagger)\widehat{U}\vec{\zeta}\,\rangle = \langle\widehat{U}\vec{\zeta} \mid \widehat{U}\,\widehat{A}\vec{\zeta}\,\rangle = \langle\vec{\zeta} \mid \widehat{A}\vec{\zeta}\,\rangle.$$

Q13(21)[4] Let $\widehat{B}$ a selfadjoint operator and let $\widehat{B}'$ be the unitary transform of $\widehat{B}$ generated by a unitary operator $\widehat{U}$. Show that $\widehat{B}$ and $\widehat{B}'$ possess the same set of eigenvalues and that their corresponding eigenvectors are unitary transforms of each other.[5]

---

[4]This question is on the important property stated in P13.4.2(6).

[5]First show that $\widehat{B}'$ possesses the same eigenvalues as $\widehat{B}$ and that the eigenvectors $\widehat{B}'$ are the corresponding unitary transforms of the eigenvectors of $\widehat{B}$. Then show that $\widehat{B}$ possesses the same eigenvalues as $\widehat{B}'$ and that the eigenvectors $\widehat{B}$ are the corresponding unitary transforms of the eigenvectors of $\widehat{B}'$.

SQ13(21) Let $\vec{\eta}$ be an eigenvector of $\widehat{B}$ corresponding to the eigenvalue b, i.e., $\widehat{B}\vec{\eta} = b\vec{\eta}$. Let $\vec{\eta}\,'$ and $\widehat{B}'$ be the unitary transforms of $\vec{\eta}$ and $\widehat{B}$ generated by the unitary operator $\widehat{U}$. Then

$$\widehat{B}'\vec{\eta}\,' = (\widehat{U}\widehat{B}\widehat{U}^\dagger)\widehat{U}\vec{\eta} = \widehat{U}\widehat{B}\vec{\eta} = b\widehat{U}\vec{\eta} = b\vec{\eta}\,',$$

showing that an eigenvalue of $\widehat{B}$ is also an eigenvalue of $\widehat{B}'$ with the corresponding unitary transform $\vec{\eta}\,'$ as eigenvector.

Conversely the operator $\widehat{B}$ can be regarded as the unitary transform of $\widehat{B}'$ generated by the unitary operator $\widehat{U}' = \widehat{U}^\dagger$. It follows that an eigenvalue b' of $\widehat{B}'$ corresponding to an eigenvector $\vec{\eta}\,'$ must also be an eigenvalue of $\widehat{B}$ with the corresponding unitary transform $\widehat{U}'\vec{\eta}\,'$ as eigenvector.

The conclusion is that $\widehat{B}$ and $\widehat{B}'$ possess the same set of eigenvalues and that their corresponding eigenvectors are unitary transforms of each other.

Q13(22) Let $\widehat{U}(t)$ be a continuous one-parameter group of unitary operators. Let $\widehat{A}$ be the generator of $\widehat{U}(t)$ in accordance with Theorem 13.4.3(2) of Stone. Let $\vec{\xi}(0)$ be an eigenvector of $\widehat{A}$ corresponding to a non-denegerate eigenvalue a. Show that

$$\widehat{U}(t)\vec{\xi}(0) = e^{-\not{i}at}\vec{\xi}(0),$$

and that

$$\vec{\xi}(t) = \widehat{U}(t)\vec{\xi}(0),$$

is a solution of Eq. (13.100).

SQ13(22) From Theorem 13.4.3(2) of Stone we know that

$$\widehat{U}(t) = e^{-\not{i}\widehat{A}t}, \quad \not{i} = \frac{i}{\hbar},$$

showing that $\widehat{U}$ is an exponential function of $\widehat{A}$. This function can be expressed in the form of Eq. (13.48), i.e.,

$$\widehat{U}(t) = \sum_{\ell=1}^{N} e^{-\not{i}a_\ell t}\,\widehat{P}_{\vec{\varepsilon}_\ell}.$$

Now choose $a_1 = a$ and $\vec{\varepsilon}_1 = \vec{\xi}(0)$ in the above equation.[6] Then we have $\widehat{P}_{\vec{\varepsilon}_1}\vec{\xi}(0) = \vec{\xi}(0)$ and $\widehat{P}_{\vec{\varepsilon}_\ell}\vec{\xi}(0) = \vec{0}$ for all $\ell \neq 1$. It follows that

$$\widehat{U}(t)\vec{\xi}(0) = \sum_{\ell=1}^{N} e^{-\not{i}a_\ell t}\,\widehat{P}_{\vec{\varepsilon}_\ell}\vec{\xi}(0) = e^{-\not{i}at}\vec{\xi}(0).$$

[6] Assuming $\vec{\xi}(0)$ is normalised. If not we can insert a normalisation constant to normalise it.

Next we have

$$\begin{aligned} i\hbar \frac{d\vec{\xi}(t)}{dt} &= i\hbar \frac{d\big(\widehat{U}(t)\vec{\xi}(0)\big)}{dt} = i\hbar \frac{d\big(e^{-iat}\vec{\xi}(0)\big)}{dt} \\ &= a\left(e^{-iat}\vec{\xi}(0)\right) = \left(e^{-iat}\widehat{A}\,\vec{\xi}(0)\right) \\ &= \left(e^{-iat}\widehat{A}\,\vec{\xi}(0)\right) = \widehat{A}\left(e^{-iat}\vec{\xi}(0)\right) \\ &= \widehat{A}\,\vec{\xi}(t). \end{aligned}$$

This shows that $\vec{\xi}(t)$ is a solution of Eq. (13.100).

The question assumes that $\widehat{A}$ has a nondegenerate eigenvalue a. A similar proof can be carried out for an eigenvector corresponding to a degenerate eigenvalue. We will then use the expression of $\widehat{U}$ in Eq. (13.49).

Chapter 14

Model Theories Based on Complex Vector Spaces

Q14(1) Verify the matrix representation of $\boldsymbol{M}_{\hat{S}_x}$ in Eq. (14.30) and of $\boldsymbol{M}_{\hat{S}_y}$ in Eq. (14.41).

SQ14(1) The matrix elements are defined by Eq. (13.107). Using Eq. (14.25) we express $\vec{\alpha}_z$ and $\vec{\beta}_z$ in terms of $\vec{\alpha}_x$ and $\vec{\beta}_x$, i.e.,

$$\vec{\alpha}_z = \frac{1}{\sqrt{2}}(\vec{\alpha}_x + \vec{\beta}_x), \quad \vec{\beta}_z = \frac{1}{\sqrt{2}}(\vec{\alpha}_x - \vec{\beta}_x).$$

Then we have

$$(\boldsymbol{M}_{\hat{S}_x})_{11} = \langle \vec{\alpha}_z \mid \widehat{S}_x \vec{\alpha}_z \rangle = \frac{1}{2} \langle \vec{\alpha}_x + \vec{\beta}_x \mid \frac{\hbar}{2}(\vec{\alpha}_x - \vec{\beta}_x) \rangle = 0.$$

$$(\boldsymbol{M}_{\hat{S}_x})_{12} = \langle \vec{\alpha}_z \mid \widehat{S}_x \vec{\beta}_z \rangle = \frac{1}{2} \langle \vec{\alpha}_x + \vec{\beta}_x \mid \frac{\hbar}{2}(\vec{\alpha}_x + \vec{\beta}_x) \rangle = \frac{\hbar}{2}.$$

$$(\boldsymbol{M}_{\hat{S}_x})_{21} = \langle \vec{\beta}_z \mid \widehat{S}_x \vec{\alpha}_z \rangle = \frac{1}{2} \langle \vec{\alpha}_x - \vec{\beta}_x \mid \frac{\hbar}{2}(\vec{\alpha}_x - \vec{\beta}_x) \rangle = \frac{\hbar}{2}.$$

$$(\boldsymbol{M}_{\hat{S}_x})_{22} = \langle \vec{\beta}_z \mid \widehat{S}_x \vec{\beta}_z \rangle = \frac{1}{2} \langle \vec{\alpha}_x - \vec{\beta}_x \mid \frac{\hbar}{2}(\vec{\alpha}_x + \vec{\beta}_x) \rangle = 0.$$

These are the matrix elements of $\boldsymbol{M}_{\hat{S}_x}$ in Eq. (14.30).

Using Eq. (14.37) we express $\vec{\alpha}_z$ and $\vec{\beta}_z$ in terms of $\vec{\alpha}_y$ and $\vec{\beta}_y$, i.e.,

$$\vec{\alpha}_z = \frac{1}{\sqrt{2}}(\vec{\alpha}_y + \vec{\beta}_y), \quad \vec{\beta}_z = -\frac{i}{\sqrt{2}}(\vec{\alpha}_y - \vec{\beta}_y).$$

Quantum Mechanics: Problems and Solutions
K. Kong Wan

ISBN 978-981-4800-72-3 (Paperback), 978-0-429-29647-5 (eBook)
www.jennystanford.com

Then we have

$$\big(\boldsymbol{M}_{\hat{S}_y}\big)_{11} = \langle \vec{\alpha}_z \mid \widehat{S}_y \vec{\alpha}_z \rangle = \frac{1}{2}\,\langle \vec{\alpha}_y + \vec{\beta}_y \mid \frac{\hbar}{2}(\vec{\alpha}_y - \vec{\beta}_y)\rangle = 0.$$

$$\big(\boldsymbol{M}_{\hat{S}_y}\big)_{12} = \langle \vec{\alpha}_z \mid \widehat{S}_y \vec{\beta}_z \rangle = -\frac{i}{2}\,\langle \vec{\alpha}_y + \vec{\beta}_y \mid \frac{\hbar}{2}(\vec{\alpha}_y + \vec{\beta}_y)\rangle = -i\,\frac{\hbar}{2}.$$

$$\big(\boldsymbol{M}_{\hat{S}_y}\big)_{21} = \langle \vec{\beta}_z \mid \widehat{S}_y \vec{\alpha}_z \rangle = \frac{i}{2}\,\langle \vec{\alpha}_y - \vec{\beta}_y \mid \frac{\hbar}{2}(\vec{\alpha}_y - \vec{\beta}_y)\rangle = i\,\frac{\hbar}{2}.$$

$$\big(\boldsymbol{M}_{\hat{S}_y}\big)_{22} = \langle \vec{\beta}_z \mid \widehat{S}_y \vec{\beta}_z \rangle = -\frac{1}{2}\,\langle \vec{\alpha}_y - \vec{\beta}_y \mid \frac{\hbar}{2}(\vec{\alpha}_y + \vec{\beta}_y)\rangle = 0.$$

Q14(2) Find the matrix representation of projectors $\widehat{P}_{\vec{\alpha}_y}$ and $\widehat{P}_{\vec{\beta}_y}$ in Eq. (14.38) in basis $\{\vec{\alpha}_z, \vec{\beta}_z\}$.

SQ14(2) Using Eq. (14.37) we express $\vec{\alpha}_y$ and $\vec{\beta}_y$ in terms of $\vec{\alpha}_z$ and $\vec{\beta}_z$, i.e.,

$$\vec{\alpha}_z = \frac{1}{\sqrt{2}}\,(\vec{\alpha}_y + \vec{\beta}_y), \quad \vec{\beta}_z = -\frac{i}{\sqrt{2}}\,(\vec{\alpha}_y - \vec{\beta}_y).$$

Then we have

$$\widehat{P}_{\vec{\alpha}_y}\vec{\alpha}_z = \langle \vec{\alpha}_y \mid \vec{\alpha}_z \rangle \vec{\alpha}_y = \frac{1}{\sqrt{2}}\,\vec{\alpha}_y,$$

$$\widehat{P}_{\vec{\alpha}_y}\vec{\beta}_z = \langle \vec{\alpha}_y \mid \vec{\beta}_z \rangle \vec{\alpha}_y = -\frac{i}{\sqrt{2}}\,\vec{\alpha}_y.$$

$$\widehat{P}_{\vec{\beta}_y}\vec{\alpha}_z = \langle \vec{\beta}_y \mid \vec{\alpha}_z \rangle \vec{\beta}_y = \frac{1}{\sqrt{2}}\,\vec{\beta}_y,$$

$$\widehat{P}_{\vec{\beta}_y}\vec{\beta}_z = \langle \vec{\beta}_y \mid \vec{\beta}_z \rangle \vec{\beta}_y = \frac{i}{\sqrt{2}}\,\vec{\beta}_y.$$

The matrix $\boldsymbol{M}_{\hat{P}_{\vec{\alpha}_y}}$ is defined by the following matrix elements:

$$\big(\boldsymbol{M}_{\hat{P}_{\vec{\alpha}_y}}\big)_{11} = \langle \vec{\alpha}_z \mid \widehat{P}_{\vec{\alpha}_y}\vec{\alpha}_z \rangle = \frac{1}{\sqrt{2}}\,\langle \vec{\alpha}_z \mid \vec{\alpha}_y \rangle = \frac{1}{2}.$$

$$\big(\boldsymbol{M}_{\hat{P}_{\vec{\alpha}_y}}\big)_{12} = \langle \vec{\alpha}_z \mid \widehat{P}_{\vec{\alpha}_y}\vec{\beta}_z \rangle = -\frac{i}{\sqrt{2}}\,\langle \vec{\alpha}_z \mid \vec{\alpha}_y \rangle = -\frac{i}{2}.$$

$$\big(\boldsymbol{M}_{\hat{P}_{\vec{\alpha}_y}}\big)_{21} = \langle \vec{\beta}_z \mid \widehat{P}_{\vec{\alpha}_y}\vec{\alpha}_z \rangle = \frac{1}{\sqrt{2}}\,\langle \vec{\beta}_z \mid \vec{\alpha}_y \rangle = \frac{i}{2}.$$

$$\big(\boldsymbol{M}_{\hat{P}_{\vec{\alpha}_y}}\big)_{22} = \langle \vec{\beta}_z \mid \widehat{P}_{\vec{\alpha}_y}\vec{\beta}_z \rangle = -\frac{i}{\sqrt{2}}\,\langle \vec{\beta}_z \mid \vec{\alpha}_y \rangle = \frac{1}{2}.$$

This is the projection matrix in Eq. (7.180).

The matrix $\boldsymbol{M}_{\hat{P}_{\vec{\beta}_y}}$ is defined by the following matrix elements:

$$\left(\boldsymbol{M}_{\hat{P}_{\vec{\beta}_y}}\right)_{11} = \langle \vec{\alpha}_z \mid \widehat{P}_{\vec{\beta}_y} \vec{\alpha}_z \rangle = \frac{1}{\sqrt{2}} \langle \vec{\alpha}_z \mid \vec{\beta}_y \rangle = \frac{1}{2}.$$

$$\left(\boldsymbol{M}_{\hat{P}_{\vec{\beta}_y}}\right)_{12} = \langle \vec{\alpha}_z \mid \widehat{P}_{\vec{\beta}_y} \vec{\beta}_z \rangle = \frac{i}{\sqrt{2}} \langle \vec{\alpha}_z \mid \vec{\beta}_y \rangle = \frac{i}{2}.$$

$$\left(\boldsymbol{M}_{\hat{P}_{\vec{\beta}_y}}\right)_{21} = \langle \vec{\beta}_z \mid \widehat{P}_{\vec{\beta}_y} \vec{\alpha}_z \rangle = \frac{1}{\sqrt{2}} \langle \vec{\beta}_z \mid \vec{\beta}_y \rangle = -\frac{i}{2}.$$

$$\left(\boldsymbol{M}_{\hat{P}_{\vec{\beta}_y}}\right)_{22} = \langle \vec{\beta}_z \mid \widehat{P}_{\vec{\beta}_y} \vec{\beta}_z \rangle = \frac{i}{\sqrt{2}} \langle \vec{\beta}_z \mid \vec{\beta}_y \rangle = \frac{1}{2}.$$

This is the projection matrix in Eq. (7.181).

Q14(3) Find the probability distribution function for the x-component spin values in the state given by the vector $\vec{\alpha}_z$. What is the corresponding expectation value?

SQ14(3) Equations (14.56) and (14.58) apply here. The state vector $\vec{\alpha}_z$ is related to the eigenvectors of $\hat{S}_x$ by Eq. (14.25), i.e., we have

$$\vec{\alpha}_x = \frac{1}{\sqrt{2}}(\vec{\alpha}_z + \vec{\beta}_z), \quad \vec{\beta}_x = \frac{1}{\sqrt{2}}(\vec{\alpha}_z - \vec{\beta}_z).$$

(1) The value of the distribution function $\mathcal{F}^{\hat{S}_x}(\vec{\alpha}_z, \tau)$ for $\tau < -\frac{1}{2}\hbar$ is zero.

(2) The value rises to 1/2 when $\tau = -\frac{1}{2}\hbar$ since

$$\wp^{\hat{S}_x}(\vec{\alpha}_z, -\frac{1}{2}\hbar) = \langle \vec{\alpha}_z \mid \widehat{P}_{\vec{\beta}_x} \vec{\alpha}_z \rangle = |\langle \vec{\alpha}_z \mid \vec{\beta}_x \rangle|^2 = \frac{1}{2}.$$

(3) The value rises to 1 when $\tau = \frac{1}{2}\hbar$ since

$$\begin{aligned}
&\wp^{\hat{S}_x}(\vec{\alpha}_z, -\frac{1}{2}\hbar) + \wp^{\hat{S}_x}(\vec{\alpha}_z, \frac{1}{2}\hbar) \\
&= \langle \vec{\alpha}_z \mid \widehat{P}_{\vec{\beta}_x} \vec{\alpha}_z \rangle + \langle \vec{\alpha}_z \mid \widehat{P}_{\vec{\alpha}_x} \vec{\alpha}_z \rangle = \langle \vec{\alpha}_z \mid (\widehat{P}_{\vec{\beta}_x} + \widehat{P}_{\vec{\alpha}_x}) \vec{\alpha}_z \rangle \\
&= \langle \vec{\alpha}_z \mid \widehat{I\!I} \vec{\alpha}_z \rangle = 1.
\end{aligned}$$

We can check this result by calculating $\wp^{\hat{S}_x}(\vec{\alpha}_z, \frac{1}{2}\hbar)$ explicitly, i.e.,

$$\wp^{\hat{S}_x}(\vec{\alpha}_z, \frac{1}{2}\hbar) = \langle \vec{\alpha}_z \mid \widehat{P}_{\vec{\alpha}_x} \vec{\alpha}_z \rangle = |\langle \vec{\alpha}_z \mid \vec{\alpha}_x \rangle|^2 = \frac{1}{2}.$$

The probability distribution function is

$$\mathcal{F}^{\hat{S}_x}(\vec{\alpha}_z, \tau) = \begin{cases} 0 & \tau < -\dfrac{\hbar}{2} \\ \dfrac{1}{2} & -\dfrac{\hbar}{2} \le \tau < \dfrac{\hbar}{2} \\ 1 & \tau \ge \dfrac{\hbar}{2} \end{cases}.$$

The expectation value is equal to

$$\left(-\frac{1}{2}\hbar\right)\wp^{\hat{S}_x}(\vec{\alpha}_z, -\frac{1}{2}\hbar) + \left(\frac{1}{2}\hbar\right)\wp^{\hat{S}_x}(\vec{\alpha}_z, \frac{1}{2}\hbar) = 0.$$

Q14(4) In the two-dimensional vector space $\vec{\mathbb{V}}^2$ we have the following matrix representations in basis $\{\vec{\alpha}_z, \vec{\beta}_z\}$:

(1) The matrix representations of the projectors $\widehat{P}_{\vec{\alpha}_x}$ and $\widehat{P}_{\vec{\beta}_x}$ are given by $\boldsymbol{M}_{\hat{P}_{\vec{\alpha}_x}}$ and $\boldsymbol{M}_{\hat{P}_{\vec{\beta}_x}}$ in Eq. (14.29).

(2) The matrix representations of projectors $\widehat{P}_{\vec{\alpha}_y}$ and $\widehat{P}_{\vec{\beta}_y}$ are given by the matrices $\boldsymbol{P}_{\vec{\alpha}_y}$ and $\boldsymbol{P}_{\vec{\beta}_y}$ in Eqs. (7.180) and (7.181). In the notation of this chapter these matrices are relabelled as $\boldsymbol{M}_{\hat{P}_{\vec{\alpha}_y}}$ and $\boldsymbol{M}_{\hat{P}_{\vec{\beta}_y}}$.

(3) The vector $\vec{\eta}$ is represented by the column vector $\boldsymbol{C}_{\vec{\eta}}$ in Eq. (14.3).

Using the expression for $\boldsymbol{M}_{\hat{P}_{\vec{\alpha}_x}} = \boldsymbol{C}_{\vec{\alpha}_x}\boldsymbol{C}^{\dagger}_{\vec{\alpha}_x}$ in Eq. (7.178) evaluate $\boldsymbol{M}_{\hat{P}_{\vec{\alpha}_x}}\boldsymbol{C}_{\vec{\eta}}$. Show that the same result is obtained using explicit matrix representations of $\boldsymbol{C}_{\vec{\alpha}_x}$ and $\boldsymbol{M}_{\hat{P}_{\vec{\alpha}_x}}$ in Eqs. (14.3), (14.26) and (14.29). Explain how the result confirms the projection nature of the matrix $\boldsymbol{M}_{\hat{P}_{\vec{\alpha}_x}}$.

Carry out a similar evaluation of $\boldsymbol{M}_{\hat{P}_{\vec{\beta}_x}}\boldsymbol{C}_{\vec{\eta}}$, $\boldsymbol{M}_{\hat{P}_{\vec{\alpha}_y}}\boldsymbol{C}_{\vec{\eta}}$ and $\boldsymbol{M}_{\hat{P}_{\vec{\beta}_y}}\boldsymbol{C}_{\vec{\eta}}$.

SQ14(4) The evaluation is straightforward. We can evaluate $\boldsymbol{M}_{\hat{P}_{\vec{\alpha}_x}}\boldsymbol{C}_{\vec{\eta}}$ in two different ways:

(1) Using Eq. (7.178) for the expression of $\boldsymbol{M}_{\hat{P}_{\vec{\alpha}_x}}$ we get

$$\boldsymbol{M}_{\hat{P}_{\vec{\alpha}_x}}\boldsymbol{C}_{\vec{\eta}} = (\boldsymbol{C}_{\vec{\alpha}_x}\boldsymbol{C}^{\dagger}_{\vec{\alpha}_x})\boldsymbol{C}_{\vec{\eta}} = \boldsymbol{C}_{\vec{\alpha}_x}(\boldsymbol{C}^{\dagger}_{\vec{\alpha}_x}\boldsymbol{C}_{\vec{\eta}}) = \langle \boldsymbol{C}_{\vec{\alpha}_x} \mid \boldsymbol{C}_{\vec{\eta}}\rangle \boldsymbol{C}_{\vec{\alpha}_x},$$

where[1]

$$\boldsymbol{C}_{\vec{\alpha}_x} = \frac{1}{\sqrt{2}}\begin{pmatrix}1\\1\end{pmatrix} \quad \text{and} \quad \langle \boldsymbol{C}_{\vec{\alpha}_x} \mid \boldsymbol{C}_{\vec{\eta}}\rangle = \frac{1}{\sqrt{2}}(c_+ + c_-).$$

[1] See Eq. (7.116) for the expression of $\boldsymbol{C}_{\vec{\alpha}_x}$.

(2) Using explicit matrix representations in Eqs. (14.3) and (14.29) and we obtain the same result, i.e.,

$$M_{\hat{P}_{\vec{\alpha}_x}} C_{\vec{\eta}} = \frac{1}{2}\begin{pmatrix} 1 & 1 \\ 1 & 1 \end{pmatrix}\begin{pmatrix} c_+ \\ c_- \end{pmatrix} = \frac{c_+ + c_-}{2}\begin{pmatrix} 1 \\ 1 \end{pmatrix}$$
$$= \frac{1}{\sqrt{2}}(c_+ + c_-) C_{\vec{\alpha}_x}.$$

The result shows that the matrix $M_{\hat{P}_{\vec{\alpha}_x}}$ projects the column vector $C_{\vec{\eta}}$ onto the unit column vector $C_{\vec{\alpha}_x}$.

Next we need to evaluate $M_{\hat{P}_{\vec{\beta}_x}} C_{\vec{\eta}}$, $M_{\hat{P}_{\vec{\alpha}_y}} C_{\vec{\eta}}$ and $M_{\hat{P}_{\vec{\beta}_y}} C_{\vec{\eta}}$.

To evaluate $M_{\hat{P}_{\vec{\beta}_x}} C_{\vec{\eta}}$ we have:

(1) Using Eq. (7.179) for the expression of $M_{\hat{P}_{\vec{\beta}_x}}$ we get

$$M_{\hat{P}_{\vec{\beta}_x}} C_{\vec{\eta}} = (C_{\vec{\beta}_x} C^{\dagger}_{\vec{\beta}_x}) C_{\vec{\eta}} = C_{\vec{\beta}_x}(C^{\dagger}_{\vec{\beta}_x} C_{\vec{\eta}})$$
$$= \langle C_{\vec{\beta}_x} \mid C_{\vec{\eta}} \rangle C_{\vec{\beta}_x},$$

where[2]

$$C_{\vec{\beta}_x} = \frac{1}{\sqrt{2}}\begin{pmatrix} 1 \\ -1 \end{pmatrix} \quad \text{and} \quad \langle C_{\vec{\beta}_x} \mid C_{\vec{\eta}} \rangle = \frac{1}{\sqrt{2}}(c_+ - c_-).$$

(2) Using explicit matrix representations in Eqs. (14.3) and (14.29) and we have

$$M_{\hat{P}_{\vec{\beta}_x}} C_{\vec{\eta}} = \frac{1}{2}\begin{pmatrix} 1 & -1 \\ -1 & 1 \end{pmatrix}\begin{pmatrix} c_+ \\ c_- \end{pmatrix} = \frac{c_+ - c_-}{2}\begin{pmatrix} 1 \\ -1 \end{pmatrix}.$$

The matrix $M_{\hat{P}_{\vec{\beta}_x}}$ projects $C_{\vec{\eta}}$ onto the unit column vector $C_{\vec{\beta}_x}$.

To evaluate $M_{\hat{P}_{\vec{\alpha}_y}} C_{\vec{\eta}}$ we have:

(1) Using Eq. (7.180) for the expression of $M_{\hat{P}_{\vec{\alpha}_y}}$ we get

$$M_{\hat{P}_{\vec{\alpha}_y}} C_{\vec{\eta}} = (C_{\vec{\alpha}_y} C^{\dagger}_{\vec{\alpha}_y}) C_{\vec{\eta}} = C_{\vec{\alpha}_y}(C^{\dagger}_{\vec{\alpha}_y} C_{\vec{\eta}})$$
$$= \langle C_{\vec{\alpha}_y} \mid C_{\vec{\eta}} \rangle C_{\vec{\alpha}_y},$$

where[3]

$$C_{\vec{\alpha}_y} = \frac{1}{\sqrt{2}}\begin{pmatrix} 1 \\ i \end{pmatrix} \quad \text{and} \quad \langle C_{\vec{\alpha}_y} \mid C_{\vec{\eta}} \rangle = \frac{1}{\sqrt{2}}(c_+ - ic_-).$$

[2] *See* Eq. (7.116) for the expression of $C_{\vec{\beta}_x}$.
[3] *See* Eq. (7.117) for the expression of $C_{\vec{\alpha}_y}$.

(2) Using explicit matrix representations in Eqs. (14.3), (14.40) and (7.180) get

$$\boldsymbol{M}_{\hat{P}_{\vec{\alpha}_x}} \boldsymbol{C}_{\vec{\eta}} = \frac{1}{2}\begin{pmatrix} 1 & -i \\ i & 1 \end{pmatrix}\begin{pmatrix} c_+ \\ c_- \end{pmatrix} = \frac{c_+ - ic_-}{2}\begin{pmatrix} 1 \\ i \end{pmatrix}.$$

The matrix $\boldsymbol{M}_{\hat{P}_{\vec{\alpha}_x}}$ projects $\boldsymbol{C}_{\vec{\eta}}$ onto the unit column vector $\boldsymbol{C}_{\vec{\alpha}_y}$.

Finally to evaluate $\boldsymbol{M}_{\hat{P}_{\vec{\beta}_y}} \boldsymbol{C}_{\vec{\eta}}$ we have:

(1) Using Eq. (7.181) for the expression of $\boldsymbol{M}_{\hat{P}_{\vec{\beta}_y}}$ we get[4]

$$\boldsymbol{M}_{\hat{P}_{\vec{\beta}_y}} \boldsymbol{C}_{\vec{\eta}} = (\boldsymbol{C}_{\vec{\beta}_y} \boldsymbol{C}^{\dagger}_{\vec{\beta}_y})\boldsymbol{C}_{\vec{\eta}} = \boldsymbol{C}_{\vec{\beta}_y}(\boldsymbol{C}^{\dagger}_{\vec{\beta}_y} \boldsymbol{C}_{\vec{\eta}}) = \langle \boldsymbol{C}_{\vec{\beta}_y} \mid \boldsymbol{C}_{\vec{\eta}} \rangle \boldsymbol{C}_{\vec{\beta}_y}.$$

where

$$\boldsymbol{C}_{\vec{\beta}_y} = \frac{1}{\sqrt{2}}\begin{pmatrix} 1 \\ -i \end{pmatrix} \quad \text{and} \quad \langle \boldsymbol{C}_{\vec{\beta}_y} \mid \boldsymbol{C}_{\vec{\eta}} \rangle = \frac{1}{\sqrt{2}}(c_+ + ic_-).$$

(2) Using Eq. (14.3) and (7.181) we have

$$\boldsymbol{M}_{\hat{P}_{\vec{\beta}_y}} \boldsymbol{C}_{\vec{\eta}} = \frac{1}{2}\begin{pmatrix} 1 & i \\ -i & 1 \end{pmatrix}\begin{pmatrix} c_+ \\ c_- \end{pmatrix} = \frac{c_+ + ic_-}{2}\begin{pmatrix} 1 \\ -i \end{pmatrix}.$$

The matrix $\boldsymbol{M}_{\hat{P}_{\vec{\beta}_y}}$ projects $\boldsymbol{C}_{\vec{\eta}}$ onto the unit column vector $\boldsymbol{C}_{\vec{\beta}_y}$.

Q14(5) Write down a formal expression of the probability density function $w^{\hat{A}}(\vec{\eta}, \tau)$ for the piecewise-constant probability distribution function $\mathcal{F}^{\hat{A}}(\vec{\eta}, \tau)$ in Eq. (14.58) in terms of Dirac delta functions.[5] Write down the Stieltjes integral in Eq. (14.60) for the expectation value in terms of the probability density function $w^{\hat{A}}(\vec{\eta}, \tau)$ obtained above and evaluate the integral.

SQ14(5) In terms of Dirac delta functions the probability density function for the distribution function in Eq. (14.58) is expressible formally as

$$\omega^{\hat{A}}(\vec{\eta}, \tau) = \frac{d\mathcal{F}^{\hat{A}}(\vec{\eta}, \tau)}{d\tau} = \wp^{\hat{A}}(\vec{\eta}, a_1)\,\delta(\tau - a_1) + \wp^{\hat{A}}(\vec{\eta}, a_2)\,\delta(\tau - a_2) + \cdots + \wp^{\hat{A}}(\vec{\eta}, a_N)\,\delta(\tau - a_N).$$

[4] *See* Eq. (7.117) for the expression of $\boldsymbol{C}_{\vec{\beta}_y}$ which agrees with Eq. (14.40).
[5] *See* Eq. (4.74).

We have used Eq. (4.74) here to express the derivative of a piecewise constant function in terms of delta functions. The expectation value is

$$\begin{aligned}\mathcal{E}(\hat{A}, \vec{\eta}) &= \int_{-\infty}^{\infty} \tau \, d\mathcal{F}^{\hat{A}}(\vec{\eta}, \tau) = \int_{-\infty}^{\infty} \tau \, \omega^{\hat{A}}(\vec{\eta}, \tau) \, d\tau \\ &= \int_{-\infty}^{\infty} \tau \left(\wp^{\hat{A}}(\vec{\eta}, a_1) \, \delta(\tau - a_1) + \wp^{\hat{A}}(\vec{\eta}, a_2) \, \delta(\tau - a_2) \right. \\ &\quad \left. + \cdots + \wp^{\hat{A}}(\vec{\eta}, a_N) \, \delta(\tau - a_N) \right) d\tau \\ &= a_1 \wp^{\hat{A}}(\vec{\eta}, a_1) + a_2 \wp^{\hat{A}}(\vec{\eta}, a_2) + \cdots + a_N \wp^{\hat{A}}(\vec{\eta}, a_N),\end{aligned}$$

as expected.

Chapter 15

Spectral Theory in Terms of Stieltjes Integrals

Q15(1) Given a spectral function $\widehat{F}(\tau)$ show that $\widehat{F}(\tau_2) - \widehat{F}(\tau_1)$ for $t_2 > t_1$ is a projector. Is $\widehat{F}(\tau_1) - \widehat{F}(\tau_2)$ also a projector?

SQ15(1) For selfadjointness we have

$$\big(\widehat{F}(\tau_2) - \widehat{F}(\tau_1)\big)^\dagger = \widehat{F}(\tau_2)^\dagger - \widehat{F}(\tau_1)^\dagger = \widehat{F}(\tau_2) - \widehat{F}(\tau_1).$$

For idempotence we have, using Eq. (15.11),

$$\begin{aligned}\big(\widehat{F}(\tau_2) - \widehat{F}(\tau_1)\big)^2 &= \widehat{F}(\tau_2)^2 + \widehat{F}(\tau_1)^2 - \widehat{F}(\tau_1)\widehat{F}(\tau_2) - \widehat{F}(\tau_2)\widehat{F}(\tau_1) \\ &= \widehat{F}(\tau_2) + \widehat{F}(\tau_1) - 2\widehat{F}(\tau_1) \\ &= \widehat{F}(\tau_2) - \widehat{F}(\tau_1).\end{aligned}$$

Hence $\widehat{F}(\tau_2) - \widehat{F}(\tau_1)$ is a projector by Theorem 13.2.2(1). If $\widehat{P}$ is a projector then $-\widehat{P}$ is not a projector since $-\widehat{P}$ is not idempotent. Similarly $\widehat{F}(\tau_1) - \widehat{F}(\tau_2) = -\big(\widehat{F}(\tau_2) - \widehat{F}(\tau_1)\big)$ is not a projector.

Q15(2) Using Eqs. (15.15), (15.19) and (15.20) prove Eqs. (15.21), (15.22) and (15.23).

Quantum Mechanics: Problems and Solutions
K. Kong Wan

ISBN 978-981-4800-72-3 (Paperback), 978-0-429-29647-5 (eBook)
www.jennystanford.com

SQ15(2)

(1) To prove Eq. (15.21) we have

$$\begin{aligned}\widehat{M}\big([\tau_1, \tau_2]\big) &= \widehat{M}\big(\{\tau_1\} \cup (\tau_1, \tau_2]\big) = \widehat{M}(\{\tau_1\}) + \widehat{M}\big((\tau_1, \tau_2]\big)\\ &= \widehat{F}(\tau_1) - \widehat{F}(\tau_1 - 0) + \widehat{F}(\tau_2) - \widehat{F}(\tau_1)\\ &= \widehat{F}(\tau_2) - \widehat{F}(\tau_1 - 0).\end{aligned}$$

(2) To prove Eq. (15.22) we have

$$\begin{aligned}\widehat{M}\big([\tau_1, \tau_2]\big) &= \widehat{M}\big([\tau_1, \tau_2) \cup \{\tau_2\}\big)\\ &= \widehat{M}\big([\tau_1, \tau_2)\big) + \widehat{M}(\{\tau_2\}).\\ \widehat{M}\big([\tau_1, \tau_2)\big) &= \widehat{M}\big([\tau_1, \tau_2]\big) - \widehat{M}(\{\tau_2\}) = \big(\widehat{F}(\tau_2) - \widehat{F}(\tau_1 - 0)\big)\\ &\quad - \big(\widehat{F}(\tau_2) - \widehat{F}(\tau_2 - 0)\big)\\ &= \widehat{F}(\tau_2 - 0) - \widehat{F}(\tau_1 - 0).\end{aligned}$$

(3) To prove Eq. (15.23) we have

$$\begin{aligned}\widehat{M}\big([\tau_1, \tau_2)\big) &= \widehat{M}\big(\{\tau_1\} \cup (\tau_1, \tau_2)\big)\\ &= \widehat{M}(\{\tau_1\}) + \widehat{M}\big((\tau_1, \tau_2)\big).\\ \widehat{M}\big((\tau_1, \tau_2)\big) &= \big(\widehat{M}\big([\tau_1, \tau_2)\big) - \widehat{M}(\{\tau_1\})\\ &= \big(\widehat{F}(\tau_2 - 0) - \widehat{F}(\tau_1 - 0)\big) - \big(\widehat{F}(\tau_1) - \widehat{F}(\tau_1 - 0)\big)\\ &= \widehat{F}(\tau_2 - 0) - \widehat{F}(\tau_1).\end{aligned}$$

Q15(3) Show that $\mathcal{F}(\vec{\eta}, \tau)$ and $\mathcal{M}(\vec{\eta}, \Lambda)$ in Eq. (15.50) define a probability distribution function and a probability measure, respectively.

SQ15(3)

(1) We need to show that $\mathcal{F}(\vec{\eta}, \tau)$ in Eq. (15.50) satisfies properties MP3.6(1), MP3.6(2) and MP3.6(3) required in Definition 3.6(2).[1]

(a) MP3.6(1) is satisfied since $\widehat{F}(\tau_2) - \widehat{F}(\tau_1)$ is selfadjoint and idempotent, i.e., we have

$$\begin{aligned}\langle \vec{\eta} \mid \widehat{F}(\tau_2)\vec{\eta}\,\rangle - \langle \vec{\eta} \mid \widehat{F}(\tau_1)\vec{\eta}\,\rangle &= \langle \vec{\eta} \mid \Big(\widehat{F}(\tau_2) - \widehat{F}(\tau_1)\Big)\vec{\eta}\,\rangle\\ &= \big|\big|\Big(\widehat{F}(\tau_2) - \widehat{F}(\tau_1)\Big)\vec{\eta}\,\big|\big|^2 \geq 0.\end{aligned}$$

[1] *See* also Definition 4.3.1(1).

(b) MP3.6(2) is obviously satisfied on account of property SF15.1(3) of spectral functions given in Definition 15.1(2), i.e., $\widehat{F}(-\infty) = \widehat{0}$ and $\widehat{F}(\infty) = \widehat{I\!I}$

(c) MP3.6(3) is also satisfied due to SF15.1(3) in Definition 15.1(2).

(2) Measures are defined by Definition 4.1.2(1) and a probability measure is a measure $\mathcal{M}$ such that $\mathcal{M}(I\!R) = 1$. Properties SM15.1(1) and SM15.1(2) of spectral measures are given in Definition 15.1(3). We can show that $\langle \vec{\eta} \mid \widehat{M}(\Lambda)\vec{\eta}\,\rangle$ satisfies the properties of a measure in the following manner:

(a) The non-negativity property is obvious, i.e., we have $\langle \vec{\eta} \mid \widehat{M}(\Lambda)\vec{\eta}\,\rangle \geq 0$ since $\widehat{M}(\Lambda)$ is selfadjoint and idempotent.

(b) For the empty set we have $\langle \vec{\eta} \mid \widehat{M}(\emptyset)\vec{\eta}\,\rangle = 0$ since $\widehat{M}(\emptyset) = \widehat{0}$ from Eq. (15.12).

(c) For additivity we have, for a set of mutually disjoint Borel sets Λ_ℓ,

$$\begin{aligned} \langle \vec{\eta} \mid \widehat{M}\big(\cup_\ell \Lambda_\ell\big)\vec{\eta}\,\rangle &= \langle \vec{\eta} \mid \sum_\ell \widehat{M}(\Lambda_\ell)\vec{\eta}\,\rangle \\ &= \sum_\ell \langle \vec{\eta} \mid \widehat{M}(\Lambda_\ell)\vec{\eta}\,\rangle. \end{aligned}$$

(d) We have shown that $\langle \vec{\eta} \mid \widehat{M}(\Lambda)\vec{\eta}\,\rangle$ is a measure. It is also a probability measure since

$$\langle \vec{\eta} \mid \widehat{M}(I\!R)\vec{\eta}\,\rangle = \langle \vec{\eta} \mid \widehat{I\!I}\vec{\eta}\,\rangle = 1.$$

Chapter 16

Infinite-Dimensional Complex Vectors and Hilbert Spaces

Q16(1) Prove the inequalities in Eq. (16.4).

SQ16(1) Let $z = a + ib$ and $w = c + id$ be two complex numbers, where i is the imaginary unit and a, b, c, d are real numbers. To prove the second inequality we start with, using Eq. (10.31),

$$|z^*w|^2 = \big(|z^*|\,|w|\big)^2 = |z|^2\,|w|^2 = (a^2+b^2)(c^2+d^2),$$

$$|z|^2 + |w|^2 = (a^2+b^2) + (c^2+d^2),$$

$$\begin{aligned}\big(|z| - |w|\big)^2 &= \left(\sqrt{a^2+b^2} - \sqrt{c^2+d^2}\right)^2 \\ &= (a^2+b^2) + (c^2+d^2) - 2\sqrt{(a^2+b^2)(c^2+d^2)} \\ &= |z|^2 + |w|^2 - 2\,|z^*w|.\end{aligned}$$

Since $\big(|z| - |w|\big)^2 \geq 0$ we get

$$|z|^2 + |w|^2 \geq 2\,|z^*w|.$$

For the first inequality we start with

$$\begin{aligned}|z+w|^2 &= (z+w)^*(z+w) = |z|^2 + |w|^2 + z^*w + zw^* \\ &= |z|^2 + |w|^2 + 2 \times \text{real part of } z^*w \\ &\leq |z|^2 + |w|^2 + 2\,|z^*w| \leq |z|^2 + |w|^2 + |z|^2 + |w|^2 \\ &= 2\big(|z|^2 + |w|^2\big),\end{aligned}$$

Quantum Mechanics: Problems and Solutions
K. Kong Wan

ISBN 978-981-4800-72-3 (Paperback), 978-0-429-29647-5 (eBook)
www.jennystanford.com

since $|z^*w| \geq$ real part of z^*w, and $|z|^2 + |w|^2 \geq 2\,|z^*w|^2$.

Q16(2) Verify Eqs. (16.21), (16.22) and (16.23).

SQ16(2) First we express $\varphi_{Hn}(x)$ in Eq. (16.18) as a function of $y = \sqrt{\lambda}\,x$, i.e.,

$$\varphi_{Hn}(y/\sqrt{\lambda}) = \left(\frac{\sqrt{\lambda}}{\sqrt{\pi}\,2^n n!}\right)^{1/2} e^{-y^2/2}\,H_n(y). \qquad (*)$$

We can directly verify Eq. (16.21), i.e., $\widehat{A}\,\varphi_{H0}(y/\sqrt{\lambda}) = 0$, using the fact that $H_0(y) = 1$ in Eq. (16.14).

Next, using Eqs. $(*)$ and (16.16) we get

$$\begin{aligned}
\widehat{A}\,\varphi_{Hn}(x) &= \widehat{A}\,\varphi_{Hn}(y/\sqrt{\lambda}) \\
&= \frac{1}{\sqrt{2}}\left(y + \frac{d}{dy}\right)\left(\frac{\sqrt{\lambda}}{\sqrt{\pi}\,2^n n!}\right)^{1/2} e^{-y^2/2}\,H_n(y) \\
&= \frac{1}{\sqrt{2}}\left(\frac{\sqrt{\lambda}}{\sqrt{\pi}\,2n\,2^{(n-1)}(n-1)!}\right)^{1/2} e^{-y^2/2}\,2nH_{n-1} \\
&= \sqrt{n}\,\varphi_{H(n-1)}(y/\sqrt{\lambda}) = \sqrt{n}\,\varphi_{H(n-1)}(x).
\end{aligned}$$

This is Eq. (16.22).

Finally, for Eq. (16.23) we can carry out similar calculations using Eqs. $(*)$ and (16.17), i.e.,

$$\begin{aligned}
\widehat{A}^*\varphi_{Hn}(x) &= \frac{1}{\sqrt{2}}\left(y - \frac{d}{dy}\right)\left(\frac{\sqrt{\lambda}}{\sqrt{\pi}\,2^n n!}\right)^{1/2} e^{-y^2/2}\,H_n(y) \\
&= \frac{1}{\sqrt{2}}\left(\frac{\sqrt{\lambda}}{\sqrt{\pi}\,2^n n!}\right)^{1/2} e^{-y^2/2}\left(2y\,H_n(y) - 2nH_{n-1}(y)\right) \\
&= \frac{1}{\sqrt{2}}\left(\frac{\sqrt{\lambda}}{\sqrt{\pi}\,2^n n!}\right)^{1/2} e^{-y^2/2}\,H_{n+1}(y) \\
&= \frac{1}{\sqrt{2}}\left(\frac{\sqrt{\lambda}\,2(n+1)}{\sqrt{\pi}\,2^{(n+1)}(n+1)!}\right)^{1/2} e^{-y^2/2}\,H_{n+1}(y) \\
&= \sqrt{n+1}\,\varphi_{H(n+1)}(y/\sqrt{\lambda}) = \sqrt{n+1}\,\varphi_{H(n+1)}(x).
\end{aligned}$$

Q16(3) Prove the equality

$$||\vec{\varphi} + \vec{\phi}\,||^2 + ||\vec{\varphi} - \vec{\phi}\,||^2 = 2||\vec{\varphi}\,||^2 + 2||\vec{\phi}\,||^2.$$

SQ16(3) Simply work out the norm explicitly:

$$||\vec{\varphi}+\vec{\phi}\,||^2 = \langle\vec{\varphi}+\vec{\phi} \mid \vec{\varphi}+\vec{\phi}\,\rangle = ||\vec{\varphi}\,||^2 + ||\vec{\phi}\,||^2 + \langle\vec{\varphi} \mid \vec{\phi}\,\rangle + \langle\phi \mid \varphi\rangle.$$

$$||\vec{\varphi}-\vec{\phi}\,||^2 = \langle\vec{\varphi}-\vec{\phi} \mid \vec{\varphi}-\vec{\phi}\,\rangle = ||\vec{\varphi}\,||^2 + ||\phi\,||^2 - \langle\vec{\varphi} \mid \vec{\phi}\,\rangle - \langle\vec{\phi} \mid \vec{\varphi}\,\rangle.$$

Adding the above two equations produces the result.

Q16(4) Prove the Schwarz inequality in Eq. (16.75).

SQ16(4) The Schwarz inequality can be proved in the same way as in a finite-dimensional complex vector space shown in SQ12(3) to question Q12(3) in Exercises and Problems Chapter 12.

Q16(5) Prove the triangle inequalities in Eqs. (16.76) and (16.77).

SQ16(5) The triangle inequalities can be proved in the same way as in a finite-dimensional complex vector space shown in SQ12(4) to question Q12(4) in Exercises and Problems for Chapter 12.

Q16(6) Show that $\phi(x),\ \psi(x) \in \mathcal{L}^2(\mathbb{R}) \Rightarrow \phi(x)+\psi(x) \in \mathcal{L}^2(\mathbb{R})$ and that the integral on the right hand side of Eq. (16.28) is finite for functions in $\mathcal{L}^2(\mathbb{R})$.

SQ16(6)[1] Since

$$\begin{aligned}|\phi(x)+\psi(x)|^2 &= (\phi(x)+\psi(x))^*(\phi(x)+\psi(x))\\ &= |\phi(x)|^2 + |\psi(x)|^2 + \phi^*(x)\psi(x) + \phi(x)\psi(x)^*,\\ |\phi(x)-\psi(x)|^2 &= (\phi(x)-\psi(x))^*(\phi(x)-\psi(x))\\ &= |\phi(x)|^2 + |\psi(x)|^2 - \phi^*(x)\psi(x) - \phi(x)\psi(x)^*,\end{aligned}$$

we get, by adding the above two equations,

$$\begin{aligned}|\phi(x)+\psi(x)|^2 &= 2\Big(|\phi(x)|^2 + |\psi(x)|^2\Big) - |\phi(x)-\psi(x)|^2\\ &\le 2\Big(|\phi(x)|^2 + |\psi(x)|^2\Big).\end{aligned}$$

Since $|\phi(x)|^2$ and $|\psi(x)|^2$ are separately integrable over $\mathbb{R}$ their sum is integrable over $\mathbb{R}$. It follows that $|\phi(x)+\psi(x)|^2$ is integrable over $\mathbb{R}$ and that $\phi(x)+\psi(x)$ is a member of $\mathcal{L}^2(\mathbb{R})$.

Similarly $|\phi(x) \pm i\psi(x)|$ are square-integrable and hence are members of $\mathcal{L}^2(\mathbb{R})$.

---

[1] Fano pp. 240–241.

To prove Eq. (16.28) we start with similar calculations, i.e.,

$$|\phi(x) + i\psi(x)|^2 = (\phi(x) + i\psi(x))^* (\phi(x) + i\psi(x))$$
$$= |\phi(x)|^2 + |\psi(x)|^2 + i\phi^*(x)\psi(x) - i\phi(x)\psi(x)^*.$$
$$|\phi(x) - i\psi(x)|^2 = (\phi(x) - i\psi(x))^* (\phi(x) - i\psi(x))$$
$$= |\phi(x)|^2 + |\psi(x)|^2 - i\phi^*(x)\psi(x) + i\phi(x)\psi(x)^*.$$

Then

$$i\,|\phi(x) - i\psi(x)|^2 - i\,|\phi(x) + i\psi(x)|^2$$
$$= 2\phi^*(x)\psi(x) - 2\phi(x)\psi(x)^*.$$

From earlier calculations we get

$$|\phi(x) + \psi(x)|^2 - |\phi(x) - \psi(x)|^2$$
$$= 2\phi^*(x)\psi(x) + 2\phi(x)\psi(x)^*.$$

Finally we get,[2] by adding the above equations,

$$\phi^*(x)\psi(x) = \frac{1}{4}\Big\{|\phi(x) + \psi(x)|^2 - |\phi(x) - \psi(x)|^2$$
$$+ i\,|\phi(x) - i\psi(x)|^2 - i\,|\phi(x) + i\psi(x)|^2\Big\}.$$

Since all the terms on the right of the equality sign are integrable over $I\!R$ we conclude that $\phi(x)^*\psi(x)$ is integrable over $I\!R$.

Q16(7) Show that the sequence of vectors $\vec{f}_n$ defined by functions $f_n(x)$ in Eq. (16.60) is a Cauchy sequence in the space $\vec{C}(\Lambda)$.[3]

SQ16(7)[4] The function $f_n(x)$ has the value 0 for $x \leq 1 - 1/n$ so that in the limit as $n \to \infty$ the function has the value zero for all $x < 1$. It follows that in the limit as $n, m \to \infty$ the function $f_n(x) - f_m(x)$ also has the value zero for all $x < 1$. Given that $f_n(x)$ and $f_m(x)$ have the value 1 for all $x \geq 1$ and hence their difference $f_n(x) - f_m(x)$ has the value 0 for all $x \geq 1$ we can deduce that $\big(f_n(x) - f_m(x)\big) \to 0$ for all x as $n, m \to \infty$ and that

$$\int_0^2 |f_n(x) - f_m(x)|^2\, dx \to 0 \text{ as } n, m \to \infty.$$

[2] Fano p. 241.
[3] Fano p. 251.
[4] Fano pp. 250–251.

Hence the set of functions $f_n(x)$ defines a Cauchy sequence of vectors $\vec{f}_n$ in $\vec{C}(\Lambda)$. Note that

$$|| \vec{f}_n - \vec{f}_m ||^2 = \int_0^2 |f_n(x) - f_m(x)|^2 \, dx.$$

Q16(8) Express the spherical harmonics in Eqs. (16.64) to (16.67) in the Cartesian coordinates x, y, z which are related to the spherical coordinates r, θ, φ by Eq. (16.41).

SQ16(8) In terms of Cartesian coordinates related to the spherical coordinates by Eq. (16.41) the expression for $Y_{0,0}$ is unchanged while the others become

$$Y_{1,-1} = \sqrt{\frac{3}{8\pi}} \big(\cos\varphi - i \sin\varphi \big) \sin\theta$$

$$= \sqrt{\frac{3}{8\pi}} \big(\cos\varphi \sin\theta - i \sin\varphi \sin\theta \big)$$

$$= \sqrt{\frac{3}{8\pi}} \frac{x - iy}{r},$$

$$Y_{1,0} = \sqrt{\frac{3}{4\pi}} \frac{z}{r},$$

$$Y_{1,1} = -\sqrt{\frac{3}{8\pi}} \frac{x + iy}{r},$$

where $r = \sqrt{x^2 + y^2 + z^2}$.

Q16(9) Show that Eq. (16.81) defines a continuous linear functional on $\vec{\mathcal{H}}$.

SQ16(9) In Eq. (16.81) we have a functional $F_{\vec{\varphi}}$ on $\vec{\mathcal{H}}$ defined by $F_{\vec{\varphi}}(\vec{\phi}\,) = \langle \vec{\varphi} \mid \vec{\phi}\, \rangle$. This functional is linear since the scalar product is linear in $\vec{\phi}$.

Let $\{\vec{\phi}_n, n = 1, 2, \ldots\}$ be a sequence of vectors converging to the vector $\vec{\phi}$. Using the Schwarz inequality we get

$$|\langle \vec{\varphi} \mid \vec{\phi}\, \rangle - \langle \vec{\varphi} \mid \vec{\phi}_n \rangle| = |\langle \vec{\varphi} \mid \vec{\phi} - \vec{\phi}_n \rangle|$$

$$\leq ||\vec{\varphi}\,||\,||\vec{\phi} - \vec{\phi}_n\,|| \to 0$$

as $\vec{\phi}_n \to \vec{\phi}$. The functional $F_{\vec{\varphi}}$ is therefore continuous.

Q16(10) In the Dirac notation a scalar product is denoted by $\langle \vec{\varphi} \mid \vec{\phi}\, \rangle$. Dirac formally consider the notation as the product of two quantities: (1) $\langle \vec{\varphi} |$ called a *bra* and (2) $| \vec{\phi} \rangle$ called a *ket*. Their product forms a *bracket* $\langle \vec{\varphi} | \vec{\phi} \rangle$ which is the scalar product. Explain how we can interpret *bras* and *kets* in terms of vectors and linear functionals.[5]

SQ16(10) A ket is just a vector by definition, i.e., $| \phi \rangle := \vec{\phi}$. Since a bra $\langle \vec{\varphi} |$ operates on a ket $\vec{\phi}$ produces the scalar product $\langle \vec{\varphi} | \vec{\phi} \rangle$ which is equal to the linear functional $F_{\vec{\varphi}}$ generated by $\vec{\varphi}$, i.e., $F_{\vec{\varphi}}(\vec{\phi}) = \langle \vec{\varphi} | \vec{\phi} \rangle$ we can see that the bra $\langle \vec{\varphi} |$ is identifiable with the linear functional $F_{\vec{\varphi}}$.

Q16(11) Explain the concept of separability in a Hilbert space.

SQ16(11) An infinite-dimensional vector space does not necessarily possess a countable orthonormal basis. The concept of separability of a space is defined by the existence of a countable orthonormal basis, i.e., a Hilbert space is separable if there is a countable orthonormal basis for the space.

---

[5]Jauch p. 32.

Chapter 17

Operators in a Hilbert Space $\vec{\mathcal{H}}$

Q17(1) Show that every linear operator on a finite-dimensional Hilbert space is bounded.[1]

SQ17(1) Let $\widehat{A}$ be a linear operator in a finite-dimensional space $\vec{\mathbb{V}}^N$. Let $\{\vec{\varepsilon}_\ell, \ell = 1, 2, \cdots, N\}$ be an orthonormal basis for $\vec{\mathbb{V}}^N$. By definition $\widehat{A}$ is defined on all the basis vectors $\vec{\varepsilon}_\ell$, i.e., $\widehat{A}\vec{\varepsilon}_\ell$ are vectors in $\vec{\mathbb{V}}^N$ with a finite norm $||\widehat{A}\vec{\varepsilon}_\ell||$. Let M be the maximum of the set of norms $||\widehat{A}\vec{\varepsilon}_\ell||$. Let $\vec{u}$ be an arbitrary unit vector in $\vec{\mathbb{V}}^N$. We have $\vec{u} = \sum_\ell c_\ell \vec{\varepsilon}_\ell$. For the norm of $\widehat{A}\vec{u}$ we have, by the triangle inequality in Eq. (16.75) and Schwarz inequality in Eq. (16.74),

$$||\widehat{A}\vec{u}\,|| = ||\sum_{\ell=1}^{N} c_\ell \widehat{A}\vec{\varepsilon}_\ell|| \leq \sum_{\ell=1}^{N} |c_\ell|\, ||\widehat{A}\vec{\varepsilon}_\ell|| \leq \sum_{\ell=1}^{N} |c_\ell|\, M$$

$$= M \sum_{\ell=1}^{N} |c_\ell| \leq MN,$$

since $|c_\ell| = |\langle \vec{\varepsilon}_\ell \mid \vec{u}\,\rangle| \leq ||\vec{\varepsilon}_\ell\,||\, ||\vec{u}\,|| = 1$. It follows that $\widehat{A}$ is bounded by Definition 17.1(2).

[1]Halmos p. 177.

Quantum Mechanics: Problems and Solutions
K. Kong Wan

ISBN 978-981-4800-72-3 (Paperback), 978-0-429-29647-5 (eBook)
www.jennystanford.com

Q17(2) Show that projectors in a Hilbert space are bounded with norm 1.

SQ17(2) Let $\widehat{P}_{\vec{S}}$ be the projector onto the subspace $\vec{S}$, and let $\vec{\phi}$ be a unit vector. Since $\vec{\phi} = \vec{\phi}_{\vec{S}} + \vec{\phi}_{\vec{S}^\perp}$ and $\vec{\phi}_{\vec{S}^\perp}$ and $\vec{\phi}_{\vec{S}^\perp}$ are orthogonal we get

$$\langle \vec{\phi} \mid \vec{\phi} \rangle = \langle \vec{\phi}_{\vec{S}} \mid \vec{\phi}_{\vec{S}} \rangle + \langle \vec{\phi}_{\vec{S}^\perp} \mid \vec{\phi}_{\vec{S}^\perp} \rangle \geq \langle \vec{\phi}_{\vec{S}} \mid \vec{\phi}_{\vec{S}} \rangle$$

$$\Rightarrow \quad ||\vec{\phi}_{\vec{S}}\,|| \leq ||\vec{\phi}\,|| = 1.$$

Since $\widehat{P}_{\vec{S}}\vec{\phi} = \vec{\phi}_{\vec{S}}$ we deduce that

$$||\widehat{P}_{\vec{S}}\,\vec{\phi}\,|| \leq ||\vec{\phi}\,|| = 1 \ \ \forall \vec{\phi}.$$

The upper bounded of $||\widehat{P}_{\vec{S}}\,\vec{\phi}\,||$ is 1 for any $\vec{\phi} \in \vec{S}$. It follows that the norm of $\widehat{P}_{\vec{S}}$ is 1.

Q17(3) Are projectors invertible? Are they reducible?

SQ17(3) Projectors are not invertible since they violate Theorem 17.5(1). The projector $\widehat{P}_{\vec{S}}$ onto the subspace $\vec{S}$ would have zero projection of any vector in the orthogonal complement $\vec{S}^\perp$ of $\vec{S}$, i.e., $\vec{\phi} \in \vec{S}^\perp \ \Rightarrow \ \widehat{P}_{\vec{S}}\vec{\phi} = \vec{0}$.

The subspace $\vec{S}$ is clearly invariant under $\widehat{P}_{\vec{S}}$, i.e., $\vec{\phi} \in \vec{S} \ \Rightarrow \ \widehat{P}_{\vec{S}}\vec{\phi} \in \vec{S}$. Since $\widehat{P}_{\vec{S}}$ is selfadjoint and bounded Theorem 17.9(1) applies, i.e., $\widehat{P}_{\vec{S}}$ is reducible by $\vec{S}$.

Q17(4) Verify that the spherical harmonics given by Eqs. (16.64) to (16.67) are eigenfunctions of the operator $\widehat{L}_z(\mathcal{S}_u)$ in Eq. (17.42), i.e.,

$$-i\hbar\,\frac{\partial Y_{\ell,m_\ell}(\theta,\varphi)}{\partial\varphi} = m_\ell\hbar\, Y_{\ell,m_\ell}(\theta,\varphi),$$

or

$$\widehat{L}_z(\mathcal{S}_u)\vec{Y}_{\ell,m_\ell} = m_\ell\hbar\,\vec{Y}_{\ell,m_\ell}.$$

SQ17(4) Using the differential expression for $\widehat{L}_z$ in Eq. (17.42), i.e., $\widehat{L}_z = -i\hbar\,\partial/\partial\varphi$ we obtain the following results:

(1) For $Y_{0,0}$ in Eq. (16.63) we have $-i\hbar\,\partial Y_{0,0}/\partial\varphi = 0$, since $Y_{0,0}$ is a constant.

(2) For $Y_{1,-1}$ in Eq. (16.66) we have $-i\hbar\,\partial Y_{1,-1}/\partial\varphi = -\hbar Y_{1,-1}$. The value $-\hbar$ comes from differentiating $\exp(-i\varphi)$ in $Y_{1,-1}$.

(3) For $Y_{1,0}$ in Eq. (16.65) we have $-i\hbar\,\partial Y_{1,0}/\partial\varphi = 0$, since $Y_{1,0}$ is independent of φ.

(4) For $Y_{1,1}$ in Eq. (16.66) we have $-i\hbar\,\partial Y_{1,1}/\partial\varphi = \hbar Y_{1,1}$. The value $\hbar$ comes from differentiating $\exp(i\varphi)$ in $Y_{1,1}$.

We can conclude that the spherical harmonics in Eqs. (16.64) to (16.67) are eigenfunctions of $\widehat{L}_z$ corresponding to eigenvalues 0, $-\hbar$, 0 and $\hbar$ respectively.[2]

Q17(5) Prove properties P17.10(1), P17.10(2) and P17.10(3) for a pair of operators $\widehat{a}$ and $\widehat{a}^\dagger$ in Definitions 17.10(1) and 17.10(2).

SQ17(5) For property P17.10(1) we have

$$\langle\vec{\varphi}_1 \mid \widehat{a}\vec{\varphi}_0\rangle = 0 \ \text{ and } \ \langle\widehat{a}\vec{\varphi}_1 \mid \vec{\varphi}_0\rangle = \langle\vec{\varphi}_0 \mid \vec{\varphi}_0\rangle = 1.$$

In other words we have $\langle\vec{\varphi}_1 \mid \widehat{a}\vec{\varphi}_0\rangle \neq \langle\widehat{a}\vec{\varphi}_1 \mid \vec{\varphi}_0\rangle$. This implies that $\widehat{a}$ is not selfadjoint since Eq. (17.102) which is necessary for selfadjointness is violated. Similarly we have

$$\langle\vec{\varphi}_1 \mid \widehat{a}^\dagger\vec{\varphi}_0\rangle = \langle\vec{\varphi}_1 \mid \vec{\varphi}_1\rangle = 1 \ \text{ and}$$
$$\langle\widehat{a}^\dagger\vec{\varphi}_1 \mid \vec{\varphi}_0\rangle = \sqrt{2}\langle\vec{\varphi}_2 \mid \vec{\varphi}_0\rangle = 0.$$

Since $\langle\vec{\varphi}_1 \mid \widehat{a}^\dagger\vec{\varphi}_0\rangle \neq \langle\widehat{a}^\dagger\vec{\varphi}_1 \mid \vec{\varphi}_0\rangle$ the operator $\widehat{a}^\dagger$ is not selfadjoint.

For property P17.10(2) we have

$$\widehat{N}\vec{\varphi}_0 = \widehat{a}^\dagger\widehat{a}\,\vec{\varphi}_0 = 0\,\vec{\varphi}_0.$$
$$\widehat{N}\vec{\varphi}_n = \widehat{a}^\dagger\widehat{a}\,\vec{\varphi}_n = \widehat{a}^\dagger\,\sqrt{n}\,\vec{\varphi}_{n-1} = \sqrt{n}\,\sqrt{(n-1)+1}\;\vec{\varphi}_{(n-1)+1}$$
$$= n\vec{\varphi}_n \quad \forall n = 1, 2, 3, \cdots.$$

These results show that $\vec{\varphi}_n$ are the eigenvectors of $\widehat{N}$ corresponding to eigenvalues $n = 0, 1, 2, \cdots$.

For property P17.10(3) we have, by the definition of the operators, for $n \geq 1$,

$$\widehat{a}\widehat{a}^\dagger\,\vec{\varphi}_n = \widehat{a}\,\sqrt{n+1}\,\varphi_{n+1} = (n+1)\vec{\varphi}_n,$$
$$\widehat{a}^\dagger\widehat{a}\,\vec{\varphi}_n = \widehat{a}\,\sqrt{n}\,\vec{\varphi}_{n-1} = n\vec{\varphi}_n$$
$$\Rightarrow\ \left(\widehat{a}\widehat{a}^\dagger - \widehat{a}^\dagger\widehat{a}\right)\vec{\varphi}_n = \vec{\varphi}_n$$
$$\Rightarrow\ [\widehat{a}, \widehat{a}^\dagger] = \widehat{I\!I}$$
$$\Rightarrow\ [\widehat{a}^\dagger, \widehat{a}] = -\widehat{I\!I}.$$

[2] *See* SQ36(1) in Exercises and Problems for Chapter 36 for calculations in Cartesian coordinates.

The above results also hold $n = 0$ since

$$\left(\widehat{a}\widehat{a}^\dagger - \widehat{a}^\dagger\widehat{a}\right)\vec{\varphi}_0 = \widehat{a}\widehat{a}^\dagger\vec{\varphi}_0 = \vec{\varphi}_0.$$

Using the formula $[\widehat{A}, \widehat{B}\widehat{C}\,] = [\widehat{A}, \widehat{B}\,]\widehat{C} + \widehat{B}[\widehat{A}, \widehat{C}\,]$ and the commutation relation $[\widehat{a}, \widehat{a}^\dagger\,] = \widehat{I\!I}$ we get

$$[\,\widehat{a}, \widehat{N}\,] = [\,\widehat{a}, \widehat{a}^\dagger\widehat{a}\,] = [\,\widehat{a}, \widehat{a}^\dagger\,]\widehat{a} = \widehat{a},$$
$$[\,\widehat{a}^\dagger, \widehat{N}] = [\,\widehat{a}^\dagger, \widehat{a}^\dagger\widehat{a}] = \widehat{a}^\dagger[\,\widehat{a}^\dagger, \widehat{a}] = -\widehat{a}^\dagger.$$

Q17(6) Find the domain of the operator $\widehat{N} := \widehat{a}^\dagger\widehat{a}$.

SQ17(6) Let $\vec{\phi}$ be a vector in the domain $\vec{\mathcal{D}}(\widehat{N})$ of $\widehat{N}$. Expressing $\vec{\phi}$ as a linear combination of $\vec{\varphi}_n$, i.e., writing $\vec{\phi} = \sum_n c_n\vec{\varphi}_n$ we have, using property P17.10(2),

$$\widehat{N}\vec{\phi} = \sum_{n=o}^{\infty} c_n\widehat{N}\vec{\varphi}_n = \sum_{n=0}^{\infty} c_n\, n\vec{\varphi}_n,$$

$$||\widehat{N}\vec{\phi}\,||^2 = \sum_{n,m=0}^{\infty} c_n^* c_m\, nm\,\langle\vec{\varphi}_n \mid \vec{\varphi}_m\rangle = \sum_{n=0}^{\infty} |c_n|^2 n^2.$$

We must require $\widehat{N}\vec{\phi}$ to have a finite norm, i.e., we must have

$$||\widehat{N}\vec{\phi}\,||^2 < \infty \text{ or equivalently } \sum_0^{\infty} |\langle\vec{\varphi}_n \mid \vec{\phi}\,\rangle|^2\, n^2 \le \infty,$$

since $c_n = \langle\vec{\varphi}_n \mid \vec{\phi}\,\rangle$. It follows that the domain of $\widehat{N}$ is

$$\vec{\mathcal{D}}(\widehat{N}) = \{\vec{\phi} \in \vec{\mathcal{H}} : \sum_{n=0}^{\infty} |\langle\vec{\varphi}_n \mid \vec{\phi}\,\rangle|^2\, n^2 \le \infty\}.$$

This is consistent with Eq. (17.58), i.e.,

$$\vec{\mathcal{D}}(\widehat{N}) = \{\vec{\phi} \in \vec{\mathcal{D}}(\widehat{a}) : \widehat{a}\vec{\phi} \in \vec{\mathcal{D}}(\widehat{a}^\dagger) = \vec{\mathcal{D}}(\widehat{a})\}.$$

Q17(7) Let $\{\vec{\varphi}_n, n = 0, 1, 2, \ldots\}$ be an orthonormal basis for a given Hilbert space. Show that the vector

$$\vec{\Phi}_z = \exp\left(-\frac{1}{2}\,|z|^2\right)\sum_{n=0}^{\infty}\frac{z^n}{\sqrt{n!}}\,\vec{\varphi}_n, \quad z \in \mathbb{C},$$

is the eigenvector of the annihilation operator $\widehat{a}$ associated with the orthonormal basis defined by Eqs. (17.114) and (17.115) corresponding to the eigenvalue z. Show also that the vector $\vec{\Phi}_z$ is normalised.

Finally show that

$$\langle \vec{\Phi}_z \mid \widehat{N}\vec{\Phi}_z \rangle = | z |^2,$$

where $\widehat{N} = \widehat{a}^\dagger \widehat{a}$.

SQ17(7) First we have

$$\begin{aligned}
\widehat{a}\,\vec{\Phi}_z &= \exp\left(-\frac{1}{2} | z |^2 \right) \sum_{n=0}^{\infty} \frac{z^n}{\sqrt{n!}} \widehat{a}\vec{\varphi}_n \\
&= \exp\left(-\frac{1}{2} | z |^2 \right) \sum_{n=1}^{\infty} \frac{z^n}{\sqrt{n!}} \sqrt{n}\,\vec{\varphi}_{n-1} \\
&= \exp\left(-\frac{1}{2} | z |^2 \right) \sum_{n=1}^{\infty} z \frac{z^{n-1}}{\sqrt{(n-1)!}} \vec{\varphi}_{n-1} \\
&= z \left\{ \exp\left(-\frac{1}{2} | z |^2 \right) \sum_{n=1}^{\infty} \frac{z^{n-1}}{\sqrt{(n-1)!}} \vec{\varphi}_{n-1} \right\} \\
&= z \left\{ \exp\left(-\frac{1}{2} | z |^2 \right) \sum_{n=0}^{\infty} \frac{z^n}{\sqrt{n!}} \vec{\varphi}_n \right\} = z\vec{\Phi}_z,
\end{aligned}$$

showing that $\vec{\Phi}_z$ is an eigenvector of the annihilation operator $\widehat{a}$ corresponding to the eigenvalue z.

For normalisation we have

$$\begin{aligned}
\langle \vec{\Phi}_z \mid \vec{\Phi}_z \rangle &= \exp(-| z |^2) \sum_{n,m=0}^{\infty} \frac{(z^n)^*}{\sqrt{n!}} \frac{z^m}{\sqrt{m!}} \langle \vec{\varphi}_n \mid \vec{\varphi}_m \rangle \\
&= \exp(-| z |^2) \sum_{n,m=0}^{\infty} \frac{(z^n)^*}{\sqrt{n!}} \frac{z^m}{\sqrt{m!}} \delta_{nm} \\
&= \exp(-| z |^2) \sum_{n=0}^{\infty} \frac{(| z |^2)^n}{n!} \\
&= \exp(-| z |^2) \exp(|z|^2) = 1.
\end{aligned}$$

Using the result $\widehat{a}\,\vec{\Phi}_z = z\,\vec{\Phi}_z$ we get

$$\begin{aligned}
\langle \vec{\Phi}_z \mid \widehat{N}\vec{\Phi}_z \rangle &= \langle \vec{\Phi}_z \mid \widehat{a}^\dagger \widehat{a}\,\vec{\Phi}_z \rangle \\
&= \langle \widehat{a}\vec{\Phi}_z \mid \widehat{a}\vec{\Phi}_z \rangle = \langle z\vec{\Phi}_z \mid z\vec{\Phi}_z \rangle = z^* z\,\langle \vec{\Phi}_z \mid \vec{\Phi}_z \rangle \\
&= | z |^2.
\end{aligned}$$

We can explicitly construct of the eigenvectors of $\widehat{a}$. Let $\vec{\Psi}_z$ be an eigenvector of $\widehat{a}$ corresponding to the eigenvalue z, i.e., $\widehat{a}\vec{\Psi}_z = z\vec{\Psi}_z$. Now expand $\vec{\Psi}_z$ as a linear combination of $\vec{\varphi}_n$, i.e.,

$$\vec{\Psi}_z = \sum_{n=0}^{\infty} \langle \vec{\varphi}_n \mid \vec{\Psi}_z \rangle \vec{\varphi}_n. \qquad (*)$$

Then we have

$$\langle \vec{\varphi}_n \mid \widehat{a}\vec{\Psi}_z \rangle = \langle \widehat{a}^\dagger \vec{\varphi}_n \mid \vec{\Psi}_z \rangle = \sqrt{n+1}\, \langle \vec{\varphi}_{n+1} \mid \vec{\Psi}_z \rangle$$

$$\Rightarrow \; z \langle \vec{\varphi}_n \mid \vec{\Psi}_z \rangle = \sqrt{n+1}\, \langle \vec{\varphi}_{n+1} \mid \vec{\Psi}_z \rangle,$$

$$\Rightarrow \; \langle \vec{\varphi}_{n+1} \mid \vec{\Psi}_z \rangle = \frac{z}{\sqrt{n+1}} \langle \vec{\varphi}_n \mid \vec{\Psi}_z \rangle.$$

This is valid for any n. Explicitly we have:

(1) For $n = 0$ we get

$$\langle \vec{\varphi}_1 \mid \vec{\Psi}_z \rangle = \frac{z}{\sqrt{1}} \langle \vec{\varphi}_0 \mid \vec{\Psi}_z \rangle.$$

(2) For $n = 1$ we get

$$\langle \vec{\varphi}_2 \mid \vec{\Psi}_z \rangle = \frac{z}{\sqrt{2}} \langle \vec{\varphi}_1 \mid \vec{\Psi}_z \rangle = \frac{z^2}{\sqrt{2 \times 1}} \langle \vec{\varphi}_0 \mid \vec{\Psi}_z \rangle.$$

(3) For $n = 2$ we get

$$\begin{aligned}\langle \vec{\varphi}_3 \mid \vec{\Psi}_z \rangle = \frac{z}{\sqrt{3}} \langle \vec{\varphi}_2 \mid \vec{\Psi}_z \rangle &= \frac{z^3}{\sqrt{3 \times 2 \times 1}} \langle \vec{\varphi}_0 \mid \vec{\Psi}_z \rangle \\ &= \frac{z^3}{\sqrt{3!}} \langle \vec{\varphi}_0 \mid \vec{\Psi}_z \rangle.\end{aligned}$$

(4) By repeating the process we get, for ≥ 1,

$$\begin{aligned}\langle \vec{\varphi}_n \mid \vec{\Psi}_z \rangle &= \frac{z}{\sqrt{n}} \langle \vec{\varphi}_{n-1} \mid \vec{\Psi}_z \rangle \\ &= \frac{z}{\sqrt{n}} \frac{z}{\sqrt{n-1}} \frac{z}{\sqrt{n-2}} \cdots \frac{z}{\sqrt{1}} \langle \vec{\varphi}_0 \mid \vec{\Psi}_z \rangle \\ &= \frac{z^n}{\sqrt{n!}} \langle \vec{\varphi}_0 \mid \vec{\Psi}_z \rangle.\end{aligned}$$

It follows that Eq. (∗) becomes

$$\vec{\Psi}_z = \sum_{n=0}^{\infty} \frac{z^n}{\sqrt{n!}} \langle \vec{\varphi}_0 \mid \vec{\Psi}_z \rangle \, \vec{\varphi}_n = \langle \vec{\varphi}_0 \mid \vec{\Psi}_z \rangle \sum_{n=0}^{\infty} \frac{z^n}{\sqrt{n!}} \vec{\varphi}_n .$$

By identifying $\langle \vec{\varphi}_0 \mid \vec{\Psi}_z \rangle$ with the normalisation constant $e^{-|z|^2/2}$ this eigenvector agrees with $\vec{\Phi}_z$.

To make the proof of the statement

$$\langle \vec{\varphi}_n \mid \vec{\Psi}_z \rangle = \frac{z^n}{\sqrt{n!}} \langle \vec{\varphi}_0 \mid \vec{\Psi}_z \rangle$$

more rigorous for all n we can use the method of (mathematical) induction. This is a standard method to prove a sequence of statement S_n to be true. The method goes as follows[3]:

(1) Show that the first statement S_1 is true.
(2) Show that if the nth statement is true then the $(n+1)^{\text{th}}$ statement must also be true, i.e.,

$$S_n \ \text{is true} \quad \Rightarrow \quad S_{n+1} \ \text{is true}.$$

(3) It follows that S_2 is true, since S_1 is true, and then S_3 is true, since S_2 is true, and so on. We can then conclude that the statement S_n is true for all n.

In the present case statement S_1 is the expression for $\langle \vec{\varphi}_1 \mid \vec{\Psi}_z \rangle$ given in item (1) and statement S_n is the expression for $\langle \vec{\varphi}_n \mid \vec{\Psi}_z \rangle$ given in item (4) earlier. We have already shown that S_1 is true, i.e.,

$$\langle \vec{\varphi}_1 \mid \vec{\Psi}_z \rangle = z \langle \vec{\varphi}_0 \mid \vec{\Psi}_z \rangle = \frac{z}{\sqrt{1!}} \langle \vec{\varphi}_0 \mid \vec{\Psi}_z \rangle .$$

Next assume that statement S_n is true, i.e., the following equation

$$\langle \vec{\varphi}_n \mid \vec{\Psi}_z \rangle = \frac{z^n}{\sqrt{n!}} \langle \vec{\varphi}_0 \mid \vec{\Psi}_z \rangle$$

is true for any given $n > 1$. Then we have

$$\begin{aligned} \langle \vec{\varphi}_{n+1} \mid \vec{\Psi}_z \rangle &= \frac{1}{\sqrt{n+1}} \langle \hat{a}^\dagger \vec{\varphi}_n \mid \vec{\Psi}_z \rangle = \frac{1}{\sqrt{n+1}} \langle \vec{\varphi}_n \mid \hat{a} \vec{\Psi}_z \rangle \\ &= \frac{z}{\sqrt{n+1}} \langle \vec{\varphi}_n \mid \vec{\Psi}_z \rangle \\ &= \frac{z}{\sqrt{(n+1)}} \left(\frac{z^n}{\sqrt{n!}} \langle \vec{\varphi}_0 \mid \vec{\Psi}_z \rangle \right) \\ &= \frac{z^{n+1}}{\sqrt{(n+1)!}} \langle \vec{\varphi}_0 \mid \vec{\Psi}_z \rangle . \end{aligned}$$

[3] *See* SQ27(8) and SQ27(13) for further examples.

In other words the statement is then true for $n + 1$. We can now conclude that the expression is true for all n.

Chapter 18

Bounded Operators on $\vec{\mathcal{H}}$

Q18(1) The characteristic function $\chi_\Lambda(x)$ of the interval Λ in $I\!R$ defines a multiplication operator $\widehat{\chi}_\Lambda$ on $\vec{L}^2(I\!R)$ by Eq. (17.11). Show that $\widehat{\chi}_\Lambda$ is a projector. Find the eigenvalues and their corresponding eigenvectors of $\widehat{\chi}_\Lambda$.

SQ18(1) The multiplication operator $\widehat{\chi}_\Lambda$ defined by the characteristic function $\chi_\Lambda(x)$ is clearly selfadjoint since $\chi_\Lambda(x)$ is real-valued so that we have, for all $\vec{\phi}, \vec{\psi} \in \vec{L}^2(I\!R)$,

$$\begin{aligned}\langle \vec{\phi} \mid \widehat{\chi}_\Lambda \vec{\psi} \rangle &= \int_{-\infty}^{\infty} \phi(x)^* \chi_\Lambda(x) \psi(x)\, dx \\ &= \int_{-\infty}^{\infty} \big(\chi_\Lambda(x)\phi(x)\big)^* \psi(x)\, dx = \langle \widehat{\chi}_\Lambda \vec{\phi} \mid \vec{\psi} \rangle.\end{aligned}$$

It is also idempotent since

$$(\widehat{\chi}_\Lambda)^2 \vec{\psi} = \widehat{\chi}_\Lambda (\widehat{\chi}_\Lambda \vec{\psi}) := (\chi_\Lambda(x))^2 \psi(x) = \chi_\Lambda(x)\psi(x).$$

The operator $\widehat{\chi}_\Lambda$ is therefore projector.

As is the case for all projectors it has two eigenvalues, 1 and 0. The eigenvectors corresponding to the eigenvalue 1 are vectors defined by functions in $L^2(I\!R)$ which vanish outside Λ while the eigenvectors corresponding to the eigenvalue 0 are vectors defined by functions in $L^2(I\!R)$ which vanish inside Λ.

Quantum Mechanics: Problems and Solutions
K. Kong Wan

ISBN 978-981-4800-72-3 (Paperback), 978-0-429-29647-5 (eBook)
www.jennystanford.com

Q18(2) Show that the multiplication operator $\widehat{x}(\Lambda)$ on $\vec{L}^2(\Lambda)$ in Eq. (17.22) is bounded and selfadjoint.

SQ18(2) Let M be the upper bound of x^2 for x in the interval Λ. Then we have, for every unit vector $\vec{\phi} \in \vec{L}^2(\Lambda)$,

$$||\widehat{x}(\Lambda)\vec{\phi}\,||^2 = \int_\Lambda x^2 \,|\phi(x)|^2 \,dx \le \int_\Lambda M^2 \,|\phi(x)|^2 \,dx = M.$$

Hence the operator is bounded. This is not true if Λ is an unbounded interval.

As a multiplication operator $\widehat{x}$ is selfadjoint since the operator satisfies the selfadjointness condition in Eq. (18.3), i.e., for any $\vec{\psi}$ and $\vec{\phi}$ in $\vec{L}^2(\Lambda)$ we have

$$\langle \vec{\psi} \mid \widehat{x}(\Lambda)\vec{\phi}\,\rangle = \int_\Lambda \psi^*(x) x\, \phi(x)\, dx = \langle \widehat{x}(\Lambda)\vec{\psi} \mid \vec{\phi}\,\rangle.$$

Q18(3) Show that plane waves $f_p(x)$ in Eq. (18.12) is not square-integrable over $I\!R$.

SQ18(3) The plane wave $f_p(x)$ has an absolutely value of $1/\sqrt{2\pi\hbar}$ for all x, i.e., $|f_p(x)| = 1/\sqrt{2\pi\hbar}$ and $|f_p(x)|^2 = 1/2\pi\hbar$. Hence the function is not integrable over $I\!R$ since

$$\int_{-\infty}^{\infty} |f_p(x)|^2 \,dx = \frac{1}{2\pi\hbar} \int_{-\infty}^{\infty} dx = \infty.$$

Q18(4) Show that

(1) The trace of a projector is equal to the dimension of the subspace onto which the projector projects.
(2) If a density operator is a projector then it is a one-dimensional projector.

SQ18(4)
(1) Let $\widehat{P}_{\vec{S}}$ be a projector onto the subspace $\vec{S}$ of a Hilbert space $\vec{\mathcal{H}}$. Let $\vec{\varphi}_m$ be an orthonormal basis for $\vec{S}$ and let $\vec{\psi}_n$ be an orthonormal basis for $\vec{S}^\perp$. Then the set of vectors $\{\vec{\varphi}_m, \vec{\psi}_n\}$ form an orthonormal basis for $\vec{\mathcal{H}}$. In accordance of Eq. (18.14) the trace of $\widehat{P}_{\vec{S}}$ is given by

$$\begin{aligned} \text{tr}(\widehat{P}_{\vec{S}}) &= \sum_m \langle \vec{\varphi}_m \mid \widehat{P}_{\vec{S}}\, \vec{\varphi}_m\rangle + \sum_n \langle \vec{\psi}_n \mid \widehat{P}_{\vec{S}}\, \vec{\psi}_n\rangle \\ &= \sum_m \langle \vec{\varphi}_m \mid \widehat{P}_{\vec{S}}\, \vec{\varphi}_m\rangle, \end{aligned}$$

using the fact that $\widehat{P}_{\vec{S}}\,\vec{\psi}_n = \vec{0}$. Since $\langle \vec{\varphi}_m \mid \widehat{P}_{\vec{S}}\,\vec{\varphi}_m\rangle = \langle \vec{\varphi}_m \mid \vec{\varphi}_m\rangle = 1$ for all $\vec{\varphi}_m$ the final sum above is equal to the dimension of the subspace $\vec{S}$.

(2) A density operator $\widehat{D}$ has trace 1. It follows that if it is also a projector then it must be a projector onto a one-dimensional subspace, i.e., it must be a one-dimensional projector.

Q18(5) Show that a convex combination of density operators as shown in Eq. (18.24) defines a density operator.

SQ18(5) First the operator is bounded since for every unit vector $\vec{\varphi}$ we have, on account of the triangle inequality in Eq. (16.76) and the result on the norm of an operator in Eq. (17.7),

$$||\widehat{D}\vec{\varphi}\,|| = ||\sum_{\ell=1}^{n} w_\ell\,\widehat{D}_\ell\vec{\varphi}\,|| \le \sum_{\ell=1}^{n} ||\,w_\ell\,\widehat{D}_\ell\vec{\varphi}\,|| = \sum_{\ell=1}^{n} w_\ell\,||\,\widehat{D}_\ell\vec{\varphi}\,||$$

$$\le \sum_{\ell=1}^{n} w_\ell\,||\,\widehat{D}_\ell\,||\,||\vec{\varphi}\,|| = \sum_{\ell=1}^{n} w_\ell\,M \le M,$$

where M is the biggest of $||\,\widehat{D}_\ell\,||$. Since all the operators involved are bounded we can use Eq. (17.95) to show that $\widehat{D}$ is selfadjoint, i.e., we have

$$\widehat{D}^\dagger = \sum_\ell w_\ell \widehat{D}_\ell^\dagger = \sum_\ell w_\ell \widehat{D}_\ell = \widehat{D}.$$

The operator is also positive since

$$\langle \vec{\phi} \mid \widehat{D}\vec{\phi}\,\rangle = \sum_{\ell=1}^{n} w_\ell \langle \vec{\phi} \mid \widehat{D}_\ell\vec{\phi}\,\rangle$$

is a sum of terms which are all positive.[1] Let $\{\vec{\varphi}_m, m = 1, 2, \ldots\}$ be an orthonormal basis for the Hilbert space. The trace of $\widehat{D}$ is

$$\mathrm{tr}\big(\widehat{D}\big) = \sum_m \langle \vec{\varphi}_m \mid \widehat{D}\vec{\varphi}_m\,\rangle = \sum_m \langle \vec{\varphi}_m \mid \sum_{\ell=1}^{n} w_\ell\,\widehat{D}_\ell\,\vec{\varphi}_m\,\rangle$$

$$= \sum_{\ell=1}^{n} w_\ell \sum_m \langle \vec{\varphi}_m \mid \widehat{D}_\ell\vec{\varphi}_m\,\rangle$$

$$= \sum_{\ell=1}^{n} w_\ell \mathrm{tr}\big(\widehat{D}_\ell\big) = \sum_{\ell=1}^{n} w_\ell = 1.$$

[1] *See* Definition 13.3.1 and P18.1(1) for definition of positive operators.

Hence $\widehat{D}$ is a density operator.

Q18(6) Show that the density operator $\widehat{D}_z$ in Eq. (18.29) can be decomposed in terms of the eigenprojectors $\widehat{P}_{\vec{\alpha}_y}$ and $\widehat{P}_{\vec{\beta}_y}$ of $\widehat{S}_y$ shown in Eq. (14.38) as

$$\widehat{D}_z = \frac{1}{2}\widehat{P}_{\vec{\alpha}_y} + \frac{1}{2}\widehat{P}_{\vec{\beta}_y}.$$

SQ18(6) Since $\widehat{P}_{\vec{\alpha}_z}$ and $\widehat{P}_{\vec{\beta}_z}$ constitute a complete orthogonal family of projectors on $\vec{\mathbb{V}}^2$ we have $\widehat{P}_{\vec{\alpha}_z} + \widehat{P}_{\vec{\beta}_z} = \widehat{\mathbb{I}}$. It follows that

$$\widehat{D}_z = \left(\widehat{P}_{\vec{\alpha}_z} + \widehat{P}_{\vec{\beta}_z}\right)/2 = \widehat{\mathbb{I}}/2.$$

Since $\widehat{P}_{\vec{\alpha}_y}$ and $\widehat{P}_{\vec{\beta}_y}$ also constitute a complete orthogonal family of projectors on $\vec{\mathbb{V}}^2$ we have $\widehat{P}_{\vec{\alpha}_y} + \widehat{P}_{\vec{\beta}_y} = \widehat{\mathbb{I}}$. It follows that

$$\widehat{D}_z = \widehat{\mathbb{I}}/2 = \left(\widehat{P}_{\vec{\alpha}_y} + \widehat{P}_{\vec{\beta}_y}\right)/2.$$

Q18(7) Find the matrix representations (density matrices) in basis $\{\vec{\alpha}_z, \vec{\beta}_z\}$ of the density operators $\widehat{D}_z$ and $\widehat{D}_x$ in Eqs. (18.29) and (18.30).

SQ18(7) Using the formula in Eq. (13.107) the matrix elements of the density matrix $\boldsymbol{D}_z$ for $\widehat{D}_z$ are easily worked out to be

$$\begin{aligned}
(\boldsymbol{D}_z)_{11} &= \langle \vec{\alpha}_z \mid \widehat{D}_z \vec{\alpha}_z \rangle = \langle \vec{\alpha}_z \mid \frac{1}{2}\left(\widehat{P}_{\vec{\alpha}_z} + \widehat{P}_{\vec{\beta}_z}\right)\vec{\alpha}_z \rangle = \frac{1}{2}, \\
(\boldsymbol{D}_z)_{12} &= \langle \vec{\alpha}_z \mid \widehat{D}_z \vec{\beta}_z \rangle = \langle \vec{\alpha}_z \mid \frac{1}{2}\left(\widehat{P}_{\vec{\alpha}_z} + \widehat{P}_{\vec{\beta}_z}\right)\vec{\beta}_z \rangle = 0, \\
(\boldsymbol{D}_z)_{21} &= \langle \vec{\beta}_z \mid \widehat{D}_z \vec{\alpha}_z \rangle = \langle \vec{\beta}_z \mid \frac{1}{2}\left(\widehat{P}_{\vec{\alpha}_z} + \widehat{P}_{\vec{\beta}_z}\right)\vec{\alpha}_z \rangle = 0, \\
(\boldsymbol{D}_z)_{22} &= \langle \vec{\beta}_z \mid \widehat{D}_z \vec{\beta}_z \rangle = \langle \vec{\beta}_z \mid \frac{1}{2}\left(\widehat{P}_{\vec{\alpha}_z} + \widehat{P}_{\vec{\beta}_z}\right)\vec{\beta}_z \rangle = \frac{1}{2}.
\end{aligned}$$

The desired density matrix is

$$\boldsymbol{D}_z = \begin{pmatrix} 1/2 & 0 \\ 0 & 1/2 \end{pmatrix} = \frac{1}{2}\begin{pmatrix} 1 & 0 \\ 0 & 1 \end{pmatrix}.$$

The density matrix $\boldsymbol{D}_x$ for $\widehat{D}_x$ can be similarly worked out. Since $\widehat{D}_x = \widehat{D}_z$ by Eq. (18.32) the same density matrix is obtained, i.e., we have

$$\boldsymbol{D}_x = \frac{1}{2}\begin{pmatrix} 1 & 0 \\ 0 & 1 \end{pmatrix}.$$

Q18(8) Show that unitary operators on a Hilbert space are invertible.

SQ18(8) For a unitary operator $\widehat{U}$ we have $\langle \widehat{U}\vec{\phi} \mid \widehat{U}\vec{\phi}\,\rangle = \langle \vec{\phi} \mid \vec{\phi}\,\rangle$. It follows that $\widehat{U}\vec{\phi} = \vec{0}$ implies $\langle \vec{\phi} \mid \vec{\phi}\,\rangle = 0$. This in turn implies $\vec{\phi} = \vec{0}$. Theorem 17.5(1) then tells us that $\widehat{U}$ is invertible.

Q18(9) Show that a unitary transformation preserves the trace of the product operator in Eq. (18.15), i.e.,

$$\text{tr}\big(\widehat{B}\widehat{D}\big) = \text{tr}\big(\widehat{B}'\widehat{D}'\big),$$

where $\widehat{B}'$ and $\widehat{D}'$ are the unitary transforms of $\widehat{B}$ and $\widehat{D}$ respectively generated by a unitary operator $\widehat{U}$ in accordance with Definition 13.4.2(1).

SQ18(9) Let $\{\vec{\varphi}_\ell\}$ be an orthonormal basis. Then

$$\text{tr}\,\big(\widehat{B}\widehat{D}\big) = \sum_\ell \langle \vec{\varphi}_\ell \mid \big(\widehat{B}\widehat{D}\big)\vec{\varphi}_\ell\,\rangle.$$

We also have

$$\begin{aligned}\text{tr}\,\big(\widehat{B}'\widehat{D}'\big) &= \sum_\ell \langle \vec{\varphi}_\ell \mid \big(\widehat{B}'\widehat{D}'\big)\vec{\varphi}_\ell\,\rangle = \sum_\ell \langle \vec{\varphi}_\ell \mid \big(\widehat{U}\,\widehat{B}\widehat{U}^{\dagger}\widehat{U}\,\widehat{D}\widehat{U}^{\dagger}\big)\vec{\varphi}_\ell\,\rangle \\ &= \sum_\ell \langle \widehat{U}^{\dagger}\vec{\varphi}_\ell \mid \widehat{B}\widehat{D}\widehat{U}^{\dagger}\vec{\varphi}_\ell\,\rangle = \sum_\ell \langle \vec{\varphi}\,'_\ell \mid \widehat{B}\widehat{D}\vec{\varphi}\,'_\ell\,\rangle,\end{aligned}$$

where $\vec{\varphi}\,'_\ell = \widehat{U}^{\dagger}\,\vec{\varphi}_\ell$. These new vectors, being unitary transforms of the basis vectors of an orthonormal basis, form a new orthonormal basis. Since the trace is independent of the choice of basis, we get

$$\sum_\ell \langle \vec{\varphi}\,'_\ell \mid \widehat{B}\widehat{D}\vec{\varphi}\,'_\ell\,\rangle = \text{tr}\,\big(\widehat{B}\widehat{D}\big) \quad\Rightarrow\quad \text{tr}\,\big(\widehat{B}'\widehat{D}'\big) = \text{tr}\,\big(\widehat{B}\widehat{D}\big).$$

Q18(10) Show that the unitary transform of a one-dimensional projector is again a one-dimensional projector.

SQ18(10) Let $\widehat{P}$ be a one-dimensional projector, i.e., $\widehat{P} = |\vec{\phi}\,\rangle\langle\vec{\phi}\,|$ for some unit vector $\vec{\phi}$. Let $\vec{\phi}\,'$ and $\widehat{P}'$ be the unitary transforms of $\vec{\phi}$ and $\widehat{P}$ generated by a unitary operator $\widehat{U}$, i.e., $\vec{\phi}\,' = \widehat{U}\vec{\phi}$ and $\widehat{P}' = \widehat{U}\,\widehat{P}\widehat{U}^{\dagger}$. We have, using Eq. (18.40),

$$|\vec{\phi}\,'\rangle\langle\vec{\phi}\,'\,| = |\widehat{U}\,\vec{\phi}\,\rangle\langle\widehat{U}\,\vec{\phi}\,| = \widehat{U}\,|\vec{\phi}\,\rangle\langle\vec{\phi}\,|\,\widehat{U}^{\dagger} = \widehat{P}'.$$

In other words $\widehat{P}'$ is the projector generated by the unit vector $\vec{\phi}\,'$. It follows that $\widehat{P}'$ is one-dimensional.

Q18(11) Verify Eq. (18.53).

SQ18(11) Using the properties of Dirac delta functions shown in Eqs. (18.44) and (18.47) we get

$$
\begin{aligned}
&\int_{-\infty}^{\infty} \underset{\sim}{\varphi}(p)^* \underset{\sim}{\varphi}(p)\, dp \\
&= \int_{-\infty}^{\infty} \left(\int_{-\infty}^{\infty} f_p^*(x)\, \phi(x)\, dx \right)^* \left(\int_{-\infty}^{\infty} f_p^*(x')\, \phi(x')\, dx' \right) dp \\
&= \int_{-\infty}^{\infty} \int_{-\infty}^{\infty} \phi^*(x)\phi(x') \left(\int_{-\infty}^{\infty} f_p(x)\, f_p^*(x')\, dp \right) dx dx' \\
&= \int_{-\infty}^{\infty} \int_{-\infty}^{\infty} \phi^*(x)\phi(x')\, \delta(x - x')\, dx dx' \\
&= \int_{-\infty}^{\infty} \phi^*(x)\phi(x)\, dx.
\end{aligned}
$$

Q18(12) Using the expression $i\hbar d/dp$ and p for $\widehat{\underset{\sim}{x}}(\underset{\sim}{I\!R})$ and $\widehat{\underset{\sim}{p}}(\underset{\sim}{I\!R})$ verify the commutation relation in Eq. (18.69).

SQ18(12)

$$
[\widehat{x}(I\!R), \widehat{p}(I\!R)] = -i\hbar\left(x\frac{d}{dx} - \frac{d}{dx}x \right) = -i\hbar\left(x\frac{d}{dx} - \left(1 + x\frac{d}{dx}\right)\right) = i\hbar.
$$

$$
[\widehat{\underset{\sim}{x}}(\underset{\sim}{I\!R}), \widehat{\underset{\sim}{p}}(\underset{\sim}{I\!R})] = i\hbar\left(\frac{d}{dp}p - p\frac{d}{dp} \right) = i\hbar\left(\left(1 + p\frac{d}{dp}\right) - p\frac{d}{dp}\right) = i\hbar.
$$

Q18(13) The Fourier transform $\vec{\underset{\sim}{\varphi}}$ of $\vec{\varphi} \in \vec{L}^2(I\!R)$ is given by the following characteristic function on the momentum space:

$$
\underset{\sim}{\varphi}(p) = \begin{cases} 0, & p \le -p_0, \\ 1/\sqrt{2p_0}, & p \in (-p_0, p_0], \\ 0, & p \ge p_0, \end{cases}
$$

where p_0 is a real and positive constant. By performing the inverse Fourier transform show that the function $\varphi(x) \in L^2(I\!R)$ corresponding to the vector $\vec{\varphi}$ is given by

$$
\varphi(x) = \sqrt{\frac{\hbar}{\pi p_0}}\, \frac{\sin(p_0 x/\hbar)}{x}.
$$

SQ18(13) The inverse Fourier transform of $\underset{\sim}{\varphi}(p)$ is

$$\begin{aligned}
\varphi(x) &= \frac{1}{\sqrt{2\pi\hbar}} \int_{-\infty}^{\infty} e^{ixp/\hbar}\, \underset{\sim}{\varphi}(p)\, dp \\
&= \frac{1}{\sqrt{2\pi\hbar}} \int_{-p_0}^{p_0} e^{ixp/\hbar}\, \frac{1}{\sqrt{2p_0}}\, dp \\
&= \frac{1}{\sqrt{2\pi\hbar}} \frac{1}{\sqrt{2p_0}} \frac{\hbar}{ix} \left(e^{ixp_0/\hbar} - e^{-ixp_0/\hbar}\right) \\
&= \sqrt{\frac{\hbar}{\pi p_0}}\, \frac{\sin(p_0 x/\hbar)}{x}.
\end{aligned}$$

Chapter 19

Symmetric and Selfadjoint Operators in $\vec{\mathcal{H}}$

Q19(1) Show that symmetric operators generate real quadratic forms, as shown in Eq. (19.2).

SQ19(1) For a symmetric operator $\widehat{A}$ we have, for any $\vec{\phi}$ in the domain of $\widehat{A}$,

$$\mathcal{Q}(\widehat{A}, \vec{\phi}\,) = \langle \vec{\phi} \mid \widehat{A}\vec{\phi}\,\rangle = \langle \widehat{A}\vec{\phi} \mid \vec{\phi}\,\rangle = \langle \vec{\phi} \mid \widehat{A}\vec{\phi}\,\rangle^* = \mathcal{Q}(\widehat{A}, \vec{\phi}\,)^*.$$

Hence $\mathcal{Q}(\widehat{A}, \vec{\phi}\,)$ is real.

Q19(2) By evaluating $\langle \vec{\phi} \mid \widehat{N}\vec{\psi}\,\rangle$ and $\langle \widehat{N}\vec{\phi} \mid \vec{\psi}\,\rangle$ separately verify that the number operator $\widehat{N}$ in Eq. (19.5) satisfies

$$\langle \vec{\phi} \mid \widehat{N}\vec{\psi}\,\rangle = \langle \widehat{N}\vec{\phi} \mid \vec{\psi}\,\rangle \quad \forall \vec{\phi},\, \vec{\psi} \in \vec{\mathcal{D}}(\widehat{N}).$$

SQ19(2) We have, for every $\vec{\phi},\, \vec{\psi} \in \vec{\mathcal{D}}(\widehat{N})$,

$$\langle \vec{\phi} \mid \widehat{N}\vec{\psi}\,\rangle = \langle \vec{\phi} \mid \widehat{a}^\dagger \widehat{a}\,\vec{\psi}\,\rangle = \langle\, \widehat{a}\vec{\phi} \mid \widehat{a}\vec{\psi}\,\rangle.$$

$$\langle \widehat{N}\vec{\phi} \mid \vec{\psi}\,\rangle = \langle\, \widehat{a}^\dagger \widehat{a}\,\vec{\phi} \mid \vec{\psi}\,\rangle = \langle\, \widehat{a}\vec{\phi} \mid \widehat{a}\vec{\psi}\,\rangle.$$

Hence $\langle \vec{\phi} \mid \widehat{N}\vec{\psi}\,\rangle = \langle \widehat{N}\vec{\phi} \mid \vec{\psi}\,\rangle$. Note that the domain $\vec{\mathcal{D}}(\widehat{N})$ which is smaller than $\vec{\mathcal{D}}(\widehat{a}) = \vec{\mathcal{D}}(\widehat{a}^\dagger)$ due Eq. (17.58).[1] It follows that $\widehat{a}^\dagger$ and $\widehat{a}$ can act on $\vec{\phi}$ and $\vec{\psi}$ which are in $\vec{\mathcal{D}}(\widehat{N})$.

[1] $\vec{\mathcal{D}}(\widehat{N})$ is worked out in SQ17(6).

Quantum Mechanics: Problems and Solutions
K. Kong Wan

ISBN 978-981-4800-72-3 (Paperback), 978-0-429-29647-5 (eBook)
www.jennystanford.com

Q19(3) Show that the number operator $\widehat{N}$ in Eq. (19.5) is bounded below. Show also that the square of a selfadjoint operator is also bounded below.

SQ19(3) We need to show that the quadratic form $\mathcal{Q}(\widehat{N}, \vec{\phi})$ for all $\vec{\phi} \in \vec{\mathcal{D}}(\widehat{N})$ is bounded below. This is seen in the following calculation:

$$\langle \vec{\phi} \mid \widehat{N}\vec{\phi} \rangle = \langle \vec{\phi} \mid \widehat{a}^{\dagger}\widehat{a}\vec{\phi} \rangle = \langle \widehat{a}\vec{\phi} \mid \widehat{a}\vec{\phi} \rangle = ||\widehat{a}\vec{\phi}\,||^2 \geq 0.$$

Hence $\widehat{N}$ is bounded below by 0 in accordance with Definition 19.1(5).

Similarly the square of a selfadjoint operator is bounded below by 0, i.e., we have $\langle \vec{\phi} \mid \widehat{A}^2\vec{\psi}\, \rangle = \langle \widehat{A}\vec{\phi} \mid \widehat{A}\vec{\phi}\, \rangle \geq 0$.

Q19(4) Explain why $\widehat{H}_D^{\infty}(\Lambda)$ cannot act on $\vec{\varphi}_{\lambda=0,n}(\Lambda)$ in Eq. (19.34) and that $\langle \vec{\varphi}_{\lambda=0,n}(\Lambda) \mid \widehat{H}_D^{\infty}(\Lambda)\vec{\varphi}_{\lambda=0,n}(\Lambda)\rangle$ is undefined.[2]

SQ19(4) The vector $\vec{\varphi}_{\lambda=0,n}(\Lambda)$ is not in the domain of the Hamiltonian $\widehat{H}_D^{\infty}(\Lambda)$ since $\vec{\varphi}_{\lambda=0,n}(\Lambda)$ violates the Dirichlet boundary condition in Eq. (17.31). It follows that

$$\langle \vec{\varphi}_{\lambda=0,n}(\Lambda) \mid \widehat{H}_D^{\infty}(\Lambda)\vec{\varphi}_{\lambda=0,n}(\Lambda)\rangle$$

is undefined. Formal calculations without reference to this result is likely to lead to erroneous conclusions. This is demonstrated by SQ19(5) below.

Q19(5) The eigenvectors $\vec{\varphi}_{D,\ell}^{\infty}(\Lambda)$ of $\widehat{H}_D^{\infty}(\Lambda)$ in Eq. (19.43) form an orthonormal basis for $\vec{L}^2(\Lambda)$. We can expression $\vec{\varphi}_{\lambda=0,n}(\Lambda)$ in Eq. (19.34) as a linear combination of $\vec{\varphi}_{D,\ell}^{\infty}(\Lambda)$. Consider the vector $\vec{\varphi}_{\lambda=0,n=0}(\Lambda)$ which corresponds to a constant function. We have

$$\vec{\varphi}_{\lambda=0,n=0}(\Lambda) = \sum_{\ell=1}^{\infty} c_\ell\, \vec{\varphi}_{D,\ell}^{\infty}(\Lambda).$$

Evaluate the coefficients c_ℓ.

Investigate whether any one of the following two procedures would yield a meaningful value for the quadratic form $\mathcal{Q}\big(\widehat{H}_D^{\infty}(\Lambda), \vec{\varphi}_{\lambda=0,n=0}(\Lambda)\big)$ given formally by

$$\langle \vec{\varphi}_{\lambda=0,n=0}(\Lambda) \mid \widehat{H}_D^{\infty}(\Lambda)\vec{\varphi}_{\lambda=0,n=0}(\Lambda)\,\rangle. \qquad (*)$$

[2] $\vec{\varphi}_{\lambda=0,n}(\Lambda)$ are the eigenvectors of operator $\widehat{p}_{\lambda=0}(\Lambda)$ in E19.3(2).

(1) Assume that $\widehat{H}_D^{\infty}(\Lambda)\vec{\varphi}_{\lambda=0,n=0}(\Lambda)$ can be calculated by

$$\widehat{H}_D^{\infty}(\Lambda)\vec{\varphi}_{\lambda=0,n=0}(\Lambda) = \widehat{H}_D^{\infty}(\Lambda)\left(\sum_{\ell=1}^{\infty} c_\ell \vec{\varphi}_{D,\ell}^{\infty}(\Lambda)\right)$$
$$= \sum_{\ell=1}^{\infty} c_\ell \widehat{H}_D^{\infty}(\Lambda)\vec{\varphi}_{D,\ell}^{\infty}(\Lambda)$$
$$= \sum_{\ell=1}^{\infty} c_\ell E_{D,\ell}^{\infty}(\Lambda)\vec{\varphi}_{D,\ell}^{\infty}(\Lambda), \qquad (**)$$

where $E_{D,\ell}^{\infty}(\Lambda)$ are the corresponding eigenvalues of $\widehat{H}_D^{\infty}(\Lambda)$ in Eq. (19.44). Then assume that the expression in Eq. (*) can be calculated by

$$\langle \vec{\varphi}_{\lambda=0,n=0}(\Lambda) \mid \sum_{\ell=1}^{\infty} c_\ell E_{D,\ell}^{\infty}(\Lambda)\vec{\varphi}_{D,\ell}^{\infty}(\Lambda)\rangle$$
$$= \sum_{\ell=1}^{\infty} c_\ell E_{D,\ell}^{\infty}(\Lambda)\langle \vec{\varphi}_{\lambda=0,n=0}(\Lambda) \mid \vec{\varphi}_{D,\ell}^{\infty}(\Lambda)\rangle$$
$$= \sum_{\ell=1}^{\infty} |c_\ell|^2 E_{D,\ell}^{\infty}(\Lambda). \qquad (***)$$

Determine whether the above sum converges.

(2) Assume that $\widehat{H}_D^{\infty}(\Lambda)\vec{\varphi}_{\lambda=0,n=0}(\Lambda)$ can be calculated by formal differentiation using Eq. (19.42). What vector would be obtained? Is the resulting value of the expression

$$\langle \vec{\varphi}_{\lambda=0,n=0}(\Lambda) \mid \widehat{H}_D^{\infty}(\Lambda)\vec{\varphi}_{\lambda=0,n=0}(\Lambda)\rangle$$

meaningful?

SQ19(5) The coefficients are given by[3]

$$c_\ell = \langle \vec{\varphi}_{D,\ell}^{\infty}(\Lambda) \mid \vec{\varphi}_{\lambda=0,n=0}(\Lambda)\rangle = \frac{\sqrt{2}}{L}\int_0^L \sin\frac{\ell\pi x}{L}\,dx$$
$$= \frac{\sqrt{2}}{\ell\pi}\left(1-\cos\ell\pi\right) = \begin{cases} 0 & \text{if } \ell \text{ is even} \\ 2\sqrt{2}/\ell\pi & \text{if } \ell \text{ is odd} \end{cases}.$$

[3] Equation (19.34) shows that the function which defines $\vec{\varphi}_{\lambda=0,n=0}(\Lambda)$ has a constant value $1/\sqrt{L}$ for all $x \in \Lambda$, i.e., the function violates the Dirichlet boundary condition in Eq. (17.31).

(1) The sum in Eq. (***) over odd integers values of ℓ, i.e., $\ell = 1, 3, 5 \cdots$, diverges, i.e.,

$$\sum_{\ell=\text{odd}}^{\infty} |c_\ell|^2 E_{D,\ell}^{\infty}(\Lambda) = \sum_{\ell=\text{odd}}^{\infty} \left(\frac{2\sqrt{2}}{\ell\pi}\right)^2 \frac{\pi^2\hbar^2\ell^2}{2mL^2}$$
$$= \sum_{\ell=\text{odd}}^{\infty} \frac{4\hbar^2}{mL^2} = \infty.$$

It follows that the above procedure fails to produce a finite value for the expression in Eq. (*).

(2) Since $\vec{\varphi}_{\lambda=0,n=0}$ corresponds to a constant function of value $1/\sqrt{L}$ for all $x \in \Lambda$ a formal differentiation of this constant function lead to the zero function which would corresponds to the zero vector in $\vec{L}^2(\Lambda)$, i.e., we will get

$$\widehat{H}_D^{\infty}(\Lambda)\vec{\varphi}_{\lambda=0,n=0}(\Lambda) = \vec{0}(\Lambda). \qquad (****)$$

This would imply

$$\langle \vec{\varphi}_{\lambda=0,n=0}(\Lambda) \mid \widehat{H}_D^{\infty}(\Lambda)\vec{\varphi}_{\lambda=0,n=0}(\Lambda)\rangle = 0.$$

This cannot be correct since Eq. (****) above implies that $\vec{\varphi}_{\lambda=0,n=0}(\Lambda)$ is an eigenvector of $\widehat{H}_D^{\infty}(\Lambda)$ corresponding to an eigenvalue 0. But Eq. (19.44) tells us that $\widehat{H}_D^{\infty}(\Lambda)$ does not have a zero eigenvalue. Another way of looking at this is to realise that

$$\langle \vec{\varphi}_{\lambda=0,n=0}(\Lambda) \mid \widehat{H}_D^{\infty}(\Lambda)\vec{\varphi}_{\lambda=0,n=0}(\Lambda)\rangle$$

cannot be smaller than the smallest eigenvalue $E_{D,1}^{\infty}(\Lambda)$ of $\widehat{H}_D^{\infty}(\Lambda)$ given by Eq. (19.44). This is because

$$\sum_{\ell=\text{odd}}^{\infty} |c_\ell|^2 E_{D,\ell}^{\infty}(\Lambda) \geq \sum_{\ell=\text{odd}}^{\infty} |c_\ell|^2 E_{D,1}^{\infty}(\Lambda) = E_{D,1}^{\infty}(\Lambda).$$

The conclusion is that $\langle \vec{\varphi}_{\lambda=0,n}(\Lambda) \mid \widehat{H}_D^{\infty}(\Lambda)\vec{\varphi}_{\lambda=0,n}(\Lambda)\rangle$ does not have a meaningful value is due to the fact that an unbounded operator $\widehat{A}$ cannot act on a vector $\vec{\psi}$ lying outside its domain. The sum in Eq. (**) does not define a vector in $\vec{L}^2(\Lambda)$ since the sum does not give a finite value for the norm of the formal vector expression, i.e., carrying out formal calculation we get

$$\langle \sum_{\ell=1}^{\infty} c_\ell E_{D,\ell}^{\infty}(\Lambda)\vec{\varphi}_\ell^{\infty}(\Lambda) \mid \sum_{m=1}^{\infty} c_m E_{D,m}^{\infty}(\Lambda)\vec{\varphi}_m^{\infty}(\Lambda)\rangle$$

$$= \sum_{\ell,m=1}^{\infty} c_\ell^* c_m E_{D,\ell}^{\infty}(\Lambda) E_m^{\infty}(\Lambda)\delta_{\ell m} = \sum_{\ell}^{\infty} |c_\ell|^2 \left(E_{D,\ell}^{\infty}(\Lambda)\right)^2$$

$$= \sum_{\ell=\mathrm{odd}}^{\infty} |c_\ell|^2 \left(E_{D,\ell}^{\infty}(\Lambda)\right)^2 = \sum_{\ell=\mathrm{odd}}^{\infty} \left(\frac{2\sqrt{2}}{\ell\pi}\right)^2 \left(\frac{\pi^2\hbar^2\ell^2}{2mL^2}\right)^2 = \infty.$$

So the expression in Eq. (∗∗∗) would not produce a meaningful result due to the fact that Eq. (∗∗) fails to define a vector. Any attempt to get round this would fail to produce any meaning result.

Chapter 20

Spectral Theory of Selfadjoint Operators in $\vec{\mathcal{H}}$

Q20(1) Show that the spectral functions of a projector $\widehat{P}$, of the zero operator $\widehat{0}$ and of the identity operator $\widehat{I\!I}$ are given, respectively by[1]

$$\widehat{F}^{\hat{0}}(\tau) = \begin{cases} \widehat{0} & \tau < 0 \\ \widehat{I\!I} & \tau \geq 0 \end{cases}.$$

$$\widehat{F}^{\hat{I\!I}}(\tau) = \begin{cases} \widehat{0} & \tau < 1 \\ \widehat{I\!I} & \tau \geq 1 \end{cases}.$$

$$\widehat{F}^{\hat{P}}(\tau) = \begin{cases} \widehat{0} & \tau < 0 \\ \widehat{I\!I} - \widehat{P} & 0 \leq \tau < 1 \\ \widehat{I\!I} & \tau \geq 1 \end{cases}.$$

SQ20(1) We are dealing with operators with a discrete spectrum. The spectral functions of these operators are therefore of the form of Eq. (20.19), i.e., it is piecewise constant, being $\widehat{0}$ at $-\infty$ and $\widehat{I\!I}$ at ∞ and with discontinuities at the eigenvalues.

(1) The zero operator $\widehat{0}$ has one eigenvalue, i.e., 0. This means that $\widehat{F}^{\hat{0}}(\tau) = \widehat{0}$ for $\tau < 0$ and at $\tau = 0$ the spectral function would

[1]Weidmann p. 195. Wan (2006) p. 152.

Quantum Mechanics: Problems and Solutions
K. Kong Wan

ISBN 978-981-4800-72-3 (Paperback), 978-0-429-29647-5 (eBook)
www.jennystanford.com

jump to a value $\widehat{F}^{\hat{0}}(0)$. The function remains constant for all $\tau \geq 0$ as there are no eigenvalues above 0. It follows that $\widehat{F}^{\hat{0}}(0)$ is equal to its value at ∞ which is $\widehat{I\!I}$. In other words we have

$$\widehat{F}^{\hat{0}}(\tau) = \begin{cases} \widehat{0} & \tau < 0 \\ \widehat{I\!I} & \tau \geq 0 \end{cases}.$$

(2) The identity operator $\widehat{I\!I}$ has one eigenvalue, i.e., 1. This means that $\widehat{F}^{\hat{I\!I}}(\tau) = \widehat{0}$ for $\tau < 1$ and at $\tau = 1$ the spectral function would jump to a value $\widehat{F}^{\hat{I\!I}}(1)$. The function remains constant for all $\tau \geq 1$ as there are no eigenvalues above 1. It follows that $\widehat{F}^{\hat{0}}(1)$ is equal to its value at ∞ which is $\widehat{I\!I}$. In other words we have

$$\widehat{F}^{\hat{I\!I}}(\tau) = \begin{cases} \widehat{0} & \tau < 1 \\ \widehat{I\!I} & \tau \geq 1 \end{cases}.$$

(3) The projector operator $\widehat{P}$ has two eigenvalue, i.e., 0 and 1. This means that $\widehat{F}^{\hat{P}}(\tau) = \widehat{0}$ for $\tau < 0$ and at $\tau = 0$ the spectral function would jump to a value $\widehat{F}^{\hat{P}}(0)$. The function $\widehat{F}^{\hat{P}}(\tau)$ remains constant for all $\tau \geq 1$ as there are no eigenvalues above 1, i.e., $\widehat{F}^{\hat{P}}(1)$ is equal to its value at ∞ which is $\widehat{I\!I}$. So we have $\widehat{F}^{\hat{P}}(\tau) = \widehat{0}$ for $\tau < 0$ and $\widehat{F}^{\hat{P}}(\tau) = \widehat{I\!I}$ for $\tau \geq 1$.

The function $\widehat{F}^{\hat{P}}(\tau)$ for $\tau \in [0, 1)$ can be determined as follows:

(a) Being a selfadjoint operator the projector $\widehat{P}$ has a spectral decomposition in the form of Eq. (20.20). Now let the eigenvalues of $\widehat{P}$ be denoted by a_1 and a_2, i.e., $a_1 = 0$ and $a_2 = 1$, and let the eigenprojectors associated with these two eigenvalues be denoted by $\widehat{P}^{\hat{P}}(a_1)$ and $\widehat{P}^{\hat{P}}(a_2)$. These eigenprojectors are related to the spectral function $\widehat{F}^{\hat{P}}(\tau)$, i.e., we have, by Eq. (15.10),

$$\widehat{P}^{\hat{P}}(a_1) = \widehat{F}^{\hat{P}}(a_1) - \widehat{F}^{\hat{P}}(a_1 - 0),$$
$$\widehat{P}^{\hat{P}}(a_2) = \widehat{F}^{\hat{P}}(a_2) - \widehat{F}^{\hat{P}}(a_2 - 0).$$

The spectral decomposition of $\widehat{P}$ becomes[2]

$$\widehat{P} = a_1 \widehat{P}^{\hat{P}}(a_1) + a_2 \widehat{P}^{\hat{P}}(a_2) = 0\widehat{P}^{\hat{P}}(0) + 1\,\widehat{P}^{\hat{P}}(1)$$
$$= \widehat{F}^{\hat{P}}(1) - \widehat{F}^{\hat{P}}(1 - 0).$$

[2]Note that $\widehat{F}^{\hat{P}}(1-0)$ is meant to be $\widehat{F}^{\hat{P}}(a_2 - 0)$ where $a_2 = 1$.

Since $\widehat{F}^{\hat{P}}(1) = \widehat{I\!I}$ we get

$$\widehat{F}^{\hat{P}}(1-0) = \widehat{I\!I} - \widehat{P}.$$

Since $\widehat{F}^{\hat{P}}(\tau)$ is constant for $\tau \in [0, 1)$ we conclude that

$$\widehat{F}^{\hat{P}}(\tau) = \widehat{I\!I} - \widehat{P} \quad \forall \tau \in [0, 1).$$

(b) Another way to look at the problem is to note that at $\tau = 1$ the increment of the spectral function is equal to the eigenprojector associated with the eigenvalue 1. From Eqs. (20.16) and (20.19) we can see that this eigenprojector is $\widehat{P}$ itself. It follows that

$$\widehat{F}^{\hat{P}}(1) - \widehat{F}^{\hat{P}}(1-0) = \widehat{P} \Rightarrow \widehat{F}^{\hat{P}}(1-0) = \widehat{I\!I} - \widehat{P}.$$

To sum up we have

$$\widehat{F}^{\hat{P}}(\tau) = \begin{cases} 0 & \tau < 0 \\ \widehat{I\!I} - \widehat{P} & 0 \le \tau < 1 \\ \widehat{I\!I} & \tau \ge 1 \end{cases}.$$

Q20(2) Show that on an interval Λ of $I\!R$ the spectral measure $\widehat{M}^{\hat{P}}(\Lambda)$ of a projector $\widehat{P}$ is related to $\widehat{P}$ by[3]

$$\widehat{M}^{\hat{P}}(\Lambda) = \begin{cases} \widehat{I\!I} - \widehat{P} & \text{if } \Lambda = \{0\} \\ \widehat{P} & \text{if } \Lambda = \{1\} \\ \widehat{0} & \text{if } \Lambda \text{ does not contain 0 or 1} \end{cases}.$$

SQ20(2) For the singleton set $\{0\}$ and $\{1\}$ we get, by Eq. (15.20),[4]

$$\widehat{M}^{\hat{P}}(\{0\}) = \widehat{F}^{\hat{P}}(0) - \widehat{F}^{\hat{P}}(0-0) = \widehat{I\!I} - \widehat{P},$$
$$\widehat{M}^{\hat{P}}(\{1\}) = \widehat{F}^{\hat{P}}(1) - \widehat{F}^{\hat{P}}(1-0) = \widehat{P}.$$

The same results can be obtained directly from Eq. (20.19).

Finally, since the spectral function is constant over any interval Λ not containing 0 or 1, we get $\widehat{M}^{\hat{P}}(\Lambda) = \widehat{0}$ for such intervals.

Q20(3) Prove that Theorem 9.4.4(1) remains valid for a selfadjoint operator with a discrete spectrum in an infinite-dimensional Hilbert space.

---

[3]Wan (2006) p. 152. *See* Q28(2) for an application.

[4]As in SQ20(1) the notation $\widehat{F}^{\hat{P}}(0-0)$ is the same as $\widehat{F}^{\hat{P}}(a_1-0)$, $a_1 = 0$, and $\widehat{F}^{\hat{P}}(1-0)$ is the same as $\widehat{F}^{\hat{P}}(a_2-0)$, $a_2 = 1$.

SQ20(3)

(1) Let $\vec{\varphi}$ be an eigenvector associated with an eigenvalue a of a selfadjoint operator $\widehat{A}$. We have

$$\begin{aligned} \langle \vec{\varphi} \mid \widehat{A}\vec{\varphi} \rangle = \langle \widehat{A}\vec{\varphi} \mid \vec{\varphi} \rangle \quad &\Rightarrow \quad \langle \vec{\varphi} \mid a\vec{\varphi} \rangle = \langle a\vec{\varphi} \mid \vec{\varphi} \rangle \\ &\Rightarrow \quad a\langle \vec{\varphi} \mid \vec{\varphi} \rangle = a^*\langle \vec{\varphi} \mid \vec{\varphi} \rangle \\ &\Rightarrow \quad a = a^* \quad \Rightarrow \quad a \in I\!R. \end{aligned}$$

(2) Let $\vec{\varphi}_1$ and $\vec{\varphi}_2$ be two eigenvectors associated with eigenvalue a_1 and a_2 respectively. We have

$$\begin{aligned} &\langle \widehat{A}\vec{\varphi}_1 \mid \vec{\varphi}_2 \rangle = \langle \vec{\varphi}_1 \mid \widehat{A}\vec{\varphi}_2 \rangle \\ \Rightarrow \quad &\langle a_1\vec{\varphi}_1 \mid \vec{\varphi}_2 \rangle = \langle \vec{\varphi} \mid a_2\vec{\varphi}_2 \rangle \\ \Rightarrow \quad &(\, a_1 - a_2)\langle \vec{\varphi}_1 \mid \vec{\varphi}_2 \rangle = 0 \\ \Rightarrow \quad &\langle \vec{\varphi}_1 \mid \vec{\varphi}_2 \rangle = 0 \ \text{ if } a_1 \neq a_2. \end{aligned}$$

(3) When the eigenvalues are all nondegenerate the spectral decomposition of the identity in Eq. (20.20) becomes $\widehat{I\!I} = \sum_\ell \widehat{P}_{\vec{\varphi}_\ell}$ as shown in Eq. (20.21) which, as pointed out in P20.3(4), shows the completeness of the eigenvectors, i.e., the eigenvectors form an orthonormal basis.

(4) Suppose the eigenvalue a_1 is degenerate and all other eigenvalues are nondegenerate. Let $\vec{S}(a_1)$ be the eigensubspace associated with this degenerate eigenvalue a_1 with degeneracy d_1 which can be infinite. We can choose an orthonormal basis $\vec{\varphi}_{1j}$, $j = 1, 2, \cdots, d_1$, for the subspace $\vec{S}(a_1)$. These are eigenvectors of the operator corresponding to the eigenvalue a_1 and they are orthogonal to the eigenvectors $\vec{\varphi}_{\ell\neq 1}$ corresponding to eigenvalues to $a_{\ell\neq 1}$. Then the eigenprojector onto the eigensubspace $\vec{S}(a_1)$ is

$$\widehat{P}^{\hat{A}}(a_1) = \sum_{j=1}^{d_1} \widehat{P}_{\vec{\varphi}_{1j}},$$

where all the projectors $\widehat{P}_{\vec{\varphi}_{1j}}$ are one-dimensional. The spectral decomposition of the identity in Eq. (20.20) can be written as

$$\widehat{I\!I} = \widehat{P}^{\hat{A}}(a_1) + \sum_{\ell\neq 1} \widehat{P}_{\vec{\varphi}_\ell}.$$

It follows that the corresponding eigenvectors $\vec{\varphi}_{1j}$ and $\vec{\varphi}_{\ell\neq 1}$ form a complete set, i.e., they form an orthonormal basis. This set is not unique since $\vec{\varphi}_{1j}$ are not unique. The analysis can be extended to apply to an operator with many degenerate eigenvalues.

(5) The projectors generated by a complete orthonormal set of eigenvectors form a complete orthogonal family of projectors.

Q20(4) In $\vec{L}^2(\mathbb{R})$ show that for every $\phi(x) \in C_c^\infty(\mathbb{R})$ we have

$$[f(\widehat{x}), \widehat{p}(\mathbb{R})]\vec{\phi} := i\hbar \frac{df(x)}{dx}\phi(x).$$

SQ20(4)

$$\begin{aligned}
[f(\widehat{x}), \widehat{p}(\mathbb{R})]\vec{\phi} &:= -i\hbar\Big(f(x)\frac{d}{dx} - \frac{d}{dx}f(x)\Big)\phi(x) \\
&= -i\hbar\Big(f(x)\frac{d}{dx} - \Big(\frac{df(x)}{dx} + f(x)\frac{d}{dx}\Big)\Big)\phi(x) \\
&= i\hbar\frac{df(x)}{dx}\phi(x).
\end{aligned}$$

Q20(5) Show that selfadjoint operators having a discrete spectrum are reducible by their eigensubspaces.

SQ20(5) A selfadjoint operator with a discrete spectrum commutes with its eigenprojectors. Let $\widehat{P}^{\widehat{A}}(a_m)$ be the projector onto the eigensubspace $\vec{\mathcal{S}}^{\widehat{A}}(a_m)$ associated with the eigenvalue a_m of a selfadjoint operator $\widehat{A}$. Then $\widehat{P}^{\widehat{A}}(a_m)$ commutes with $\widehat{A}$. Theorem 17.9(1) then tells us that $\widehat{A}$ is reducible by $\vec{\mathcal{S}}^{\widehat{A}}(a_m)$.

Q20(6) Show that in $\vec{L}^2(\mathbb{R})$ the position operator $\widehat{x}(\mathbb{R})$ is reducible and the momentum operator $\widehat{p}(\mathbb{R})$ is also reducible.[5]

SQ20(6) Generally a selfadjoint operator commutes with its spectral projectors. For the position operator $\widehat{x}$ in $\vec{L}^2(\mathbb{R})$ its spectral projector associated with an interval Λ is given by Eq. (20.28), i.e., $\widehat{\chi}_\Lambda$. The subspace $\vec{\mathcal{S}}^{\widehat{x}}(\Lambda)$ associated with the projector is defined by all the functions in $L^2(\mathbb{R})$ vanishing outside the interval Λ. Since $\widehat{\chi}_\Lambda$ commute with $\widehat{x}$ Theorem 17.9(1) tells us $\widehat{x}$ is reducible by the

[5]This is in contrast to the fact discussed in E20.7(1) that together the position and momentum operators form an irreducible set.

subspace $\vec{S}^{\hat{x}}(\Lambda)$. The same analysis in the momentum space shows that the momentum operator is similarly reducible.

Q20(7) Show that the position operator $\hat{x}$ in $\vec{L}^2(I\!R)$ constitutes a complete set of selfadjoint operators.

SQ20(7)[6] The proof presented below is based on an application of Theorem 20.6(1), i.e., we want to show that any bounded operator on $\vec{L}^2(I\!R)$ commuting with $\hat{x}$ is a function of $\hat{x}$.

Let $\phi(x)$ be a function in $L^2(I\!R)$ and let $\psi(x)$ be a real positive function in $L^2(I\!R)$, i.e., $\psi(x) > 0$ for all x. Let $\vec{\phi}$ and $\vec{\psi}$ be the vectors in $\vec{L}^2(I\!R)$ defined by $\phi(x)$ and $\psi(x)$. We can introduce a function $G(x) = \phi(x)/\psi(x)$ since $\psi(x) > 0$ for all x. This function $G(x)$ defines a multiplication operator $\widehat{G}$ in $\vec{L}^2(I\!R)$ by

$$\widehat{G}\vec{\varphi} := \frac{\phi(x)}{\psi(x)}\,\varphi(x),$$

where the vector $\vec{\varphi} := \varphi(x) \in L^2(I\!R)$. We have

$$\vec{\phi} = \widehat{G}\vec{\psi} \quad \text{since} \quad \widehat{G}\vec{\psi} := \frac{\phi(x)}{\psi(x)}\,\psi(x) = \phi(x).$$

The operator $\widehat{G}$ is seen to be a multiplication operator defined by the function $G(x) := \phi(x)/\psi(x)$. As such a multiplication operator $\widehat{G}$ is a function of the position operator $\hat{x}$, i.e., $\widehat{G} = G(\hat{x})$. Note that once a real positive function $\psi(x) \in L^2(I\!R)$ is chosen we can define an operator $\widehat{G}$ for every function $\phi(x)$ in $L^2(I\!R)$.

Let $\widehat{B}$ be a bounded operator on $\vec{L}^2(I\!R)$ which commutes with the position operator $\hat{x}$. Then $\widehat{B}$ commutes with $\widehat{G}$ since $\widehat{G}$ is a function of $\hat{x}$. Let $\vec{\Psi} = \widehat{B}\vec{\psi}$ and let $\Psi(x)$ be the function in $L^2(I\!R)$ which corresponds to the vector $\vec{\Psi}$, i.e., $\vec{\Psi} = \widehat{B}\vec{\psi} := \Psi(x)$. Then

$$\widehat{B}\vec{\phi} = \widehat{B}\widehat{G}\vec{\psi} = \widehat{G}\big(\widehat{B}\vec{\psi}\,\big) = \widehat{G}\,\vec{\Psi} := \frac{\phi(x)}{\psi(x)}\,\Psi(x) = \frac{\Psi(x)}{\psi(x)}\,\phi(x).$$

It follows that $\widehat{B}$ is a multiplication operator corresponding to the function $f(x) = \Psi(x)/\psi(x)$. By Definition 20.6(3) the position operator constitutes a complete set of selfadjoint operators in $\vec{L}^2(I\!R)$.

---

[6] Jordan p. 58.

Chapter 21

Spectral Theory of Unitary Operators on $\vec{\mathcal{H}}$

Q21(1) Let $\widehat{U}$ be a unitary operator in the Hilbert space $\vec{\mathcal{H}}$ with a spectral decomposition given by Eq. (21.4). Let $\vec{\phi}$ be an arbitrary vector in $\vec{\mathcal{H}}$. Show that

$$\widehat{U}\vec{\phi} = \sum_{\ell} c_\ell\, e^{ia_\ell}\, \vec{\varphi}_\ell, \quad c_\ell = \langle \vec{\varphi}_\ell \mid \vec{\phi} \rangle.$$

SQ21(1) Using $\widehat{P}_{\vec{\varphi}_\ell}\vec{\phi} = c_\ell \vec{\varphi}_\ell, c_\ell = \langle \vec{\varphi}_\ell \mid \vec{\phi}\,\rangle$ we get

$$\widehat{U}\vec{\phi} = \left(\sum_{\ell=1}^{\infty} e^{ia_\ell}\, \widehat{P}_{\vec{\varphi}_\ell} \right) \vec{\phi} = \sum_{\ell=1}^{\infty} e^{ia_\ell} \left(\widehat{P}_{\vec{\varphi}_\ell} \vec{\phi} \right) = \sum_{\ell=1}^{\infty} c_\ell e^{ia_\ell} \vec{\varphi}_\ell.$$

Q21(2) Equation (10.27), which is a time dependent Schrödinger equation, can be written in the form of Eq. (21.18) which follows from Stone's theorem as a vector equation in the Hilbert space $\vec{L}^2(I\!R)$, i.e.,

$$i\hbar \frac{d\vec{\phi}(t)}{dt} = \widehat{H}\,\vec{\phi}(t),$$

where $\vec{\phi}(t)$ is in the domain of $\widehat{H}$. Explain why the norm of vector $\vec{\phi}(t)$ is preserved in time, i.e., $||\vec{\phi}(t_1)|| = ||\vec{\phi}(t_2)||$ for any t_1 and t_2.

Quantum Mechanics: Problems and Solutions
K. Kong Wan

ISBN 978-981-4800-72-3 (Paperback), 978-0-429-29647-5 (eBook)
www.jennystanford.com

SQ21(2) The solution of Eq. (21.20) for $\vec{\phi}(0) := \vec{\phi}(t = 0)$ is easily verified to be

$$\vec{\phi}(t) = \widehat{U}(t)\vec{\phi}(0), \quad \widehat{U}(t) = e^{-i\widehat{H}t},$$

since

$$i\hbar\frac{d\vec{\phi}(t)}{dt} = i\hbar\frac{d\widehat{U}(t)}{dt}\vec{\phi}(0) = \widehat{H}\vec{\phi}(t).$$

In other words $\vec{\phi}(t)$ is a unitary transform of $\vec{\phi}(0)$. A unitary transform preserves the norm of vectors, i.e., $||\vec{\phi}(t)|| = ||\vec{\phi}(0)||$.

Q21(3) Consider the following unitary transformations of an operator $\widehat{A}$:

$$\widehat{A}(t) = \widehat{U}^{\dagger}(t)\widehat{A}\widehat{U}(t),$$

where $\widehat{U}(t) = e^{-it\widehat{H}}$ is a one-parameter family of unitary operators in Eq. (21.13).[1] Show that[2]

$$i\hbar\frac{d\widehat{A}(t)}{dt} = [\widehat{A}(t), \widehat{H}\,].$$

SQ21(3)

$$\begin{aligned} i\hbar\frac{d\widehat{A}(t)}{dt} &= i\hbar\Big(\frac{d\widehat{U}^{\dagger}(t)}{dt}\,\widehat{A}\,\widehat{U}(t) + \widehat{U}^{\dagger}(t)\,\widehat{A}\,\frac{d\widehat{U}(t)}{dt}\Big) \\ &= -\widehat{H}\,\widehat{U}^{\dagger}(t)\,\widehat{A}\widehat{U}(t) + \widehat{U}^{\dagger}(t)\,\widehat{A}\widehat{H}\,\widehat{U}(t) \\ &= -\widehat{H}\,\widehat{A}(t) + \widehat{A}(t)\widehat{H} = [\widehat{A}(t), \widehat{H}\,]. \end{aligned}$$

[1]The operators $\widehat{A}$ and $\widehat{H}$ are independent of t and $\widehat{A}(t = 0) = \widehat{A}$.
[2]*See* Eq. (29.19) for the physical relevance of this result.

Chapter 22

Selfadjoint Operators, Unit Vectors and Probability Distributions

Q22(1) The commutation relation between the selfadjoint position and momentum operators $\widehat{x} = x$ and $\widehat{p} = -i\hbar\, d/dx$ in the Hilbert space $\vec{L}^2(I\!R)$ is often written as

$$[\widehat{x}, \widehat{p}\,] = i\hbar.$$

As pointed out in the discussion in §17.7 this relation should be expressed as an inequality, i.e.,

$$[\widehat{x}, \widehat{p}\,] \subset i\hbar \widehat{I\!I},$$

or

$$[\widehat{x}, \widehat{p}\,]\,\vec{\phi} = i\hbar\vec{\phi} \qquad (*)$$

for an appropriate set of vectors $\vec{\phi}$ in $\vec{L}^2(I\!R)$. What are the conditions $\vec{\phi}$ must satisfy in order for the equality to hold?

SQ22(1) What we want is the domain $\vec{\mathcal{D}}\big([\widehat{x}, \widehat{p}\,]\big)$ of the commutator $[\widehat{x}, \widehat{p}\,]$. The domains $\vec{\mathcal{D}}(\widehat{x})$ and $\vec{\mathcal{D}}(\widehat{p})$ of the position and momentum operators are given by Eqs. (17.13) and (17.49). The commutator is $\widehat{x}\widehat{p} - \widehat{p}\widehat{x}$. So we must specify the domains of $\widehat{x}\widehat{p}$

Quantum Mechanics: Problems and Solutions
K. Kong Wan

ISBN 978-981-4800-72-3 (Paperback), 978-0-429-29647-5 (eBook)
www.jennystanford.com

and $\widehat{p}\,\widehat{x}$ first in terms that of $\vec{\mathcal{D}}(\widehat{x})$ and $\vec{\mathcal{D}}(\widehat{p})$, and then specify the domain of the commutator. We have, by Eq. (17.58),

$$\vec{\mathcal{D}}(\widehat{x}\,\widehat{p}) := \{\vec{\phi} \in \vec{L}(I\!R) : \vec{\phi} \in \vec{\mathcal{D}}(\widehat{p}),\ \ \widehat{p}\,\vec{\phi} \in \vec{\mathcal{D}}(\widehat{x})\}.$$
$$\vec{\mathcal{D}}(\widehat{p}\,\widehat{x}) := \{\vec{\phi} \in \vec{L}(I\!R) : \vec{\phi} \in \vec{\mathcal{D}}(\widehat{x}),\ \ \widehat{x}\,\vec{\phi} \in \vec{\mathcal{D}}(\widehat{p})\}.$$

Finally we have, by Eq. (17.56), the domain

$$\vec{\mathcal{D}}([\widehat{x},\,\widehat{p}\,]) = \vec{\mathcal{D}}(\widehat{x}\,\widehat{p}) \cap \vec{\mathcal{D}}(\widehat{p}\,\widehat{x}).$$

Vectors $\vec{\phi}$ in Eq. (∗) must belong to $\vec{\mathcal{D}}\big([\widehat{x},\,\widehat{p}\,]\big)$.

Q22(2) Consider a particle confined in an infinite potential well of width $[0, L]$. All its wave functions $\phi(x)$ must vanish outside the well, i.e., $\phi(x) = 0\ \ \forall x \notin [0, L]$. Hence the state space of the trapped particle is taken to be $\vec{L}^2(\Lambda)$, $\Lambda = [0, L]$ rather than $\vec{L}^2(I\!R)$.

(a) Taking the operator $\widehat{x}(\Lambda)$ in Eq. (17.22) as the position operator show that the uncertainty in position cannot be bigger than the width of the well.

(b) Taking the operator $\widehat{p}_{\lambda=0}(\Lambda)$ in Eq. (17.36) as the momentum operator show that the momentum uncertainty

$$\Delta\big(\widehat{p}_{\lambda=0}(\Lambda),\, \vec{\varphi}_{\lambda=0,n}(\Lambda)\big)$$

is zero. Here $\vec{\varphi}_{\lambda=0,n}(\Lambda)$ are eigenvectors of $\widehat{p}_{\lambda=0}(\Lambda)$ in Eq. (19.34).

(c) Bearing in mind the above results investigate whether or not an uncertainty relation similar to that shown in Eq. (22.28) remains valid for eigenvectors of $\widehat{p}_{\lambda=0}(\Lambda)$.[1]

SQ22(2)(a) Let $\vec{\phi}$ be a unit vector in $\vec{L}^2(\Lambda)$ defined by a normalised function $\phi(x) \in L^2(\Lambda)$. For the infinite the potential well of width L from $x = 0$ to $x = L$ we have $x \le L$. We have

$$\langle \vec{\phi} \mid \widehat{x}(\Lambda)^2 \vec{\phi}\,\rangle = \int_0^L x^2\,|\phi(x)|^2\,dx \le L^2 \int_0^L |\phi(x)|^2\,dx = L^2.$$

It follows from Eq. (22.4) that

$$\begin{aligned}\Delta(\widehat{x}(\Lambda),\vec{\phi}\,)^2 &= \langle \vec{\phi} \mid \widehat{x}(\Lambda)^2\,\vec{\phi}\,\rangle - \langle \vec{\phi} \mid \widehat{x}(\Lambda)\vec{\phi}\,\rangle^2\\ &\le \langle \vec{\phi} \mid \widehat{x}(\Lambda)^2\vec{\phi}\,\rangle \le L^2\\ &\Rightarrow \Delta(\widehat{x}(\Lambda),\vec{\phi}\,) \le L.\end{aligned}$$

[1]Fano pp. 407–408 for a similar problem with the uncertainty relation in the Hilbert space $\vec{L}^2(C_a)$. *See* also §28.3.3 on a particle in circular motion.

SQ22(2)(b) For an eigenvector $\vec{\varphi}_{\lambda=0,n}(\Lambda)$ of $\widehat{p}_{\lambda=0}(\Lambda)$ corresponding to the eigenvalue $p_{\lambda=0,n}(\Lambda)$ in Eq. (19.35) we have

$$\langle \vec{\varphi}_{\lambda=0,n}(\Lambda) \mid \widehat{p}_{\lambda=0}(\Lambda)\vec{\varphi}_{\lambda=0,n}(\Lambda)\rangle = p_{\lambda=0,n}(\Lambda),$$
$$\langle \vec{\varphi}_{\lambda=0,n}(\Lambda) \mid \widehat{p}_{\lambda=0}(\Lambda)^2\vec{\varphi}_{\lambda=0,n}(\Lambda)\rangle = \big(p_{\lambda=0,n}(\Lambda)\big)^2.$$

It follows that

$$\begin{aligned}\Delta\big(\widehat{p}_{\lambda=0}(\Lambda\,),\vec{\varphi}_{\lambda=0,n}(\Lambda)\big) &= \langle \vec{\varphi}_{\lambda=0,n}(\Lambda) \mid \widehat{p}_{\lambda=0}(\Lambda\,)^2\vec{\varphi}_{\lambda=0,n}(\Lambda)\rangle \\ &\quad -\langle \vec{\varphi}_{\lambda=0,n}(\Lambda) \mid \widehat{p}_{\lambda=0}(\Lambda\,)\vec{\varphi}_{\lambda=0,n}(\Lambda)\rangle^2 \\ &= 0,\end{aligned}$$

a result which is intuitively obvious.

SQ22(2)(c) The results in SQ22(2)(a) and SQ22(2)(b) imply

$$\Delta\big(\widehat{x}(\Lambda\,),\vec{\varphi}_{\lambda=0,n}(\Lambda)\big)\Delta\big(\widehat{p}_{\lambda=0}(\Lambda\,),\vec{\varphi}_{\lambda=0,n}(\Lambda)\big) = 0.$$

This result violates the inequality in Eq. (22.28), i.e., the inequality cannot be applied to eigenvectors of $\widehat{p}_{\lambda=0}(\Lambda)$. The reason for this violation lies in the domain of the commutator discussed in SQ22(1), i.e., the commutator $[\,\widehat{x}(\Lambda\,),\,\widehat{p}_{\lambda=0}(\Lambda)\,]$ is not defined on the eigenvector $\vec{\varphi}_{\lambda=0,n}(\Lambda)$ in Eq. (19.34). To see this let the commutator acts on a vector $\vec{\phi} \in \vec{L}^2(\Lambda)$, i.e., we have

$$[\,\widehat{x}(\Lambda\,),\,\widehat{p}_{\lambda=0}(\Lambda)\,]\vec{\phi} = \widehat{x}(\Lambda)\,\widehat{p}_{\lambda=0}(\Lambda)\,\vec{\phi} - \widehat{p}_{\lambda=0}(\Lambda)\,\widehat{x}(\Lambda)\,\vec{\phi}.$$

For above operation to be meaningful we require that:

(1) The vector $\vec{\phi}$ must in the domain of $\widehat{p}_{\lambda=0}(\Lambda)$, i.e., apart from differentiability the corresponding function $\phi(x) \in L^2(\Lambda)$ must satisfy the periodic boundary condition $\phi(0) = \phi(L)$.
(2) The function $x\phi(x)$ corresponding to the vector $\widehat{x}(\Lambda)\vec{\phi}$ must also satisfy the periodic boundary condition.

Clearly the function $x\varphi_{\lambda=0,n}(\Lambda)(x)$ corresponding to the vector $\widehat{x}(\Lambda\,)\,\vec{\varphi}_{\lambda=0,n}(\Lambda)$, i.e.,

$$\frac{x}{\sqrt{L}}\exp\left[i\left(\frac{2n\pi}{L}\right)x\right],$$

does not satisfy the periodic boundary condition. This means that the vector $\widehat{x}(\Lambda\,)\,\vec{\varphi}_{\lambda=0,n}(\Lambda)$ is not in the domain of $\widehat{p}_{\lambda=0}(\Lambda)$. Consequently $\widehat{p}_{\lambda=0}(\Lambda)$ cannot act on $\widehat{x}(\Lambda)\,\vec{\varphi}_{\lambda=0,n}(\Lambda)$. It follows that the commutator cannot act on $\vec{\varphi}_{\lambda=0,n}(\Lambda)$, which in turn implies that the inequality in Eq. (22.28) cannot be applied to $\vec{\varphi}_{\lambda=0,n}(\Lambda)$.

A similar problem arises with the uncertainty relation in the Hilbert space $\vec{L}^2(C_a)$ for a particle in circular motion. This is discussed in §28.3.2.

Chapter 23

Physics of Unitary Transformations

Q23(1) Explain why a coordinate representation space and its corresponding momentum representation space introduced in §18.4.2 give rise to two mathematically different descriptions of the position and momentum which are physically equivalent.

SQ23(1) The coordinate and momentum representations are related by a Fourier transformation which is a unitary transformation. The two representations are therefore physically equivalent. Take the example of a particle in one-dimensional motion. The representation space is $\vec{L}^2(I\!R)$ in the coordinate representation. The position and momentum operators $\widehat{x}(I\!R)$ and $\widehat{p}(I\!R)$ are defined by Eqs. (17.12), (17.14), (17.49) and (17.50). The representation space is $\underset{\sim}{\vec{L}}^2(\underset{\sim}{I\!R})$ in the momentum representation. The position and momentum operators $\underset{\sim}{\widehat{x}}(\underset{\sim}{I\!R})$ and $\underset{\sim}{\widehat{p}}(\underset{\sim}{I\!R})$ in the momentum representation are defined by Eqs. (18.58) and (18.61). The quantities in two representation spaces are related by a Fourier transformation, i.e., from Eqs. (18.54), (18.57) and(18.60),

$$\underset{\sim}{\vec{\varphi}} = \widehat{U}_F \vec{\varphi}, \quad \vec{\varphi} \in \vec{L}^2(I\!R), \quad \underset{\sim}{\vec{\varphi}} \in \underset{\sim}{\vec{L}}^2(\underset{\sim}{I\!R}),$$

$$\underset{\sim}{\widehat{x}}(\underset{\sim}{I\!R}) = \widehat{U}_F\, \widehat{x}(I\!R)\, \widehat{U}_F^{-1}, \quad \underset{\sim}{\widehat{p}}(\underset{\sim}{I\!R}) = \widehat{U}_F\, \widehat{p}(I\!R)\, \widehat{U}_F^{-1},$$

where $\widehat{U}_F$ is the Fourier transform operator defined by Eq. (18.54). Since $\widehat{U}_F$ is unitary all the measured values of physical quantities,

Quantum Mechanics: Problems and Solutions
K. Kong Wan

ISBN 978-981-4800-72-3 (Paperback), 978-0-429-29647-5 (eBook)
www.jennystanford.com

e.g., eigenvalues, expectation values, probabilities, are the same in the two representations on account of Eq. (23.2), e.g.,

$$\mathcal{Q}(\widehat{x}(I\!R), \vec{\varphi}) = \mathcal{Q}(\widehat{\underset{\sim}{x}}(\underset{\sim}{I\!R}), \underset{\sim}{\vec{\varphi}}), \quad \mathcal{Q}(\widehat{p}(I\!R), \vec{\varphi}) = \mathcal{Q}(\widehat{\underset{\sim}{p}}(\underset{\sim}{I\!R}), \underset{\sim}{\vec{\varphi}}).$$

The two representations are therefore physically equivalent.

Chapter 24

Direct Sums and Tensor Products of Hilbert Spaces and Operators

Q24(1) Consider a three-dimensional Hilbert space $\vec{\mathcal{H}}$ with a preferred decomposition as the direct sum of a complete orthogonal family of one-dimensional subspaces $\vec{\mathcal{S}}^{(-)}$, $\vec{\mathcal{S}}^{(0)}$ and $\vec{\mathcal{S}}^{(+)}$, i.e.,

$$\vec{\mathcal{H}} = \vec{\mathcal{H}}^{\oplus} = \vec{\mathcal{S}}^{(-)} \oplus \vec{\mathcal{S}}^{(0)} \oplus \vec{\mathcal{S}}^{(+)}.$$

Let $\vec{\eta}^{(-)}$, $\vec{\eta}^{(0)}$ and $\vec{\eta}^{(+)}$ be unit vectors in $\vec{\mathcal{S}}^{(-)}$, $\vec{\mathcal{S}}^{(0)}$ and $\vec{\mathcal{S}}^{(+)}$, respectively.[1] A vector in $\vec{\mathcal{H}}^{\oplus}$ is of the form

$$\begin{aligned}\vec{\eta}^{\oplus} &= c_- \, \vec{\eta}^{(-)} \oplus c_0 \, \vec{\eta}^{(0)} \oplus c_+ \, \vec{\eta}^{(+)} \\ &= c_- \, \vec{\eta}^{(-)\oplus} + c_0 \, \vec{\eta}^{(0)\oplus} + c_+ \, \vec{\eta}^{(+)\oplus},\end{aligned}$$

where $c_-, \; c_0, \; c_+ \in \mathbb{C}$.

(a) Show that selfadjoint decomposable operators $\widehat{A}^{\oplus}$ on $\vec{\mathcal{H}}^{\oplus}$ are diagonalisable and of the form

$$\widehat{A}^{\oplus} = a_- \, \widehat{I\!I}^{(-)} \oplus a_0 \, \widehat{I\!I}^{(0)} \oplus a_+ \, \widehat{I\!I}^{(+)}, \quad a_-, a_0, a_+ \in I\!R.$$

What are the eigenvalues and eigenvectors of $\widehat{A}^{\oplus}$?

[1]The notation such as $\vec{\eta}^{(-)\oplus}$ follows that of Eq. (24.23), i.e., $\vec{\eta}^{(-)\oplus} = \vec{\eta}^{(-)} \oplus \vec{0}^{(0)} \oplus \vec{0}^{(+)}$ which is a vector in $\vec{\mathcal{H}}^{\oplus}$.

Quantum Mechanics: Problems and Solutions
K. Kong Wan

ISBN 978-981-4800-72-3 (Paperback), 978-0-429-29647-5 (eBook)
www.jennystanford.com

(b) Define three operators $\widehat{L}_-$, $\widehat{L}_+$ and $\widehat{L}$ on $\vec{\mathcal{H}}^\oplus$ by[2]

$$\widehat{L}_-\vec{\eta}^{(-)\oplus} = \vec{\eta}^{(0)\oplus}, \quad \widehat{L}_-\vec{\eta}^{(0)\oplus} = \vec{\eta}^{(-)\oplus}, \quad \widehat{L}_-\vec{\eta}^{(+)\oplus} = \vec{0}^{\oplus}.$$

$$\widehat{L}_+\vec{\eta}^{(+)\oplus} = \vec{\eta}^{(0)\oplus}, \quad \widehat{L}_+\vec{\eta}^{(0)\oplus} = \vec{\eta}^{(+)\oplus}, \quad \widehat{L}_+\vec{\eta}^{(-)\oplus} = \vec{0}^{\oplus}.$$

$$\widehat{L} = \widehat{L}_- + \widehat{L}_+.$$

Show that these operators are selfadjoint but not decomposable.[3]

SQ24(1)(a) By Definition 24.1.2(2) a selfadjoint decomposable operator $\widehat{A}^\oplus$ on $\vec{\mathcal{H}}^\oplus$ is of the form

$$\widehat{A}^\oplus = \widehat{A}^{(-)} \oplus \widehat{A}^{(0)} \oplus \widehat{A}^{(+)},$$

where $\widehat{A}^{(-)}$, $\widehat{A}^{(0)}$, $\widehat{A}^{(+)}$ are sefladjoint operators defined on the spaces $\vec{\mathcal{S}}^{(-)}$, $\vec{\mathcal{S}}^{(0)}$, $\vec{\mathcal{S}}^{(+)}$ respectively. Since these spaces are one-dimensional $\widehat{A}^{(-)}$, $\widehat{A}^{(0)}$, $\widehat{A}^{(+)}$ must be proportional to the identity operators in the spaces, i.e.,

$$\widehat{A}^{(-)} = a_-\widehat{I\!I}^{(-)}, \quad \widehat{A}^{(0)} = a_0\widehat{I\!I}^{(0)}, \quad \widehat{A}^{(+)} = a_+\widehat{I\!I}^{(+)},$$

where a_1, a_2, a_3 real numbers. The eigenvalues of $\widehat{A}^\oplus$ are a_1, a_2, a_3 corresponding respectively to eigenvectors

$$\vec{\eta}^{(-)\oplus} = \vec{\eta}^{(-)} \oplus \vec{0}^{(0)} \oplus \vec{0}^{(+)}, \quad \vec{\eta}^{(0)\oplus} = \vec{0}^{(-)} \oplus \vec{\eta}^{(0)} \oplus \vec{0}^{(+)},$$

$$\vec{\eta}^{(+)\oplus} = \vec{0}^{(-)} \oplus \vec{0}^{(0)} \oplus \vec{\eta}^{(+)}.$$

SQ24(1)(b) We can appreciate the action of the operators $\widehat{L}_-$ and $\widehat{L}_+$ as follows:

(1) $\widehat{L}_-$ interchanges $\vec{\eta}^{(-)\oplus}$ and $\vec{\eta}^{(0)\oplus}$ while annihilating $\vec{\eta}^{(+)\oplus}$, i.e., $\widehat{L}_-\vec{\eta}^{(-)\oplus} = \vec{\eta}^{(0)\oplus}$, $\widehat{L}_-\vec{\eta}^{(0)\oplus} = \vec{\eta}^{(-)\oplus}$, $\widehat{L}_-\vec{\eta}^{(+)\oplus} = \vec{0}^{\oplus}$.

(2) $\widehat{L}_+$ interchanges $\vec{\eta}^{(+)\oplus}$ and $\vec{\eta}^{(0)\oplus}$ while annihilating $\vec{\eta}^{(-)\oplus}$, i.e., $\widehat{L}_+\vec{\eta}^{(+)\oplus} = \vec{\eta}^{(0)\oplus}$, $\widehat{L}_+\vec{\eta}^{(0)\oplus} = \vec{\eta}^{(+)\oplus}$, $\widehat{L}_+\vec{\eta}^{(-)\oplus} = \vec{0}^{\oplus}$.

---

[2]Wan (2006) p. 356. These operators are not the direct sums of operators on $\vec{\mathcal{S}}^{(-)}$, $\vec{\mathcal{S}}^{(0)}$ and $\vec{\mathcal{S}}^{(+)}$. Hence they are not denoted with a superscript $\oplus$. Here $\widehat{0}^{\oplus}$ is the zero operator on $\vec{\mathcal{H}}^\oplus$.

[3]*See* §32.3 and §34.7 for physical applications and Q32(3) for a similar operator in a two-dimensional space.

We need to verify the selfadjointness condition on the basis vectors $\vec{\eta}^{(-)\oplus}$, $\vec{\eta}^{(0)\oplus}$, $\vec{\eta}^{(+)\oplus}$ in $\vec{\mathcal{H}}^{\oplus}$, e.g., for $\widehat{L}_-$ we need to verify that $\langle \vec{\eta}^{(\ell)\oplus} \mid \widehat{L}_- \vec{\eta}^{(\ell')\oplus} \rangle = \langle \widehat{L}_- \vec{\eta}^{(\ell)\oplus} \mid \vec{\eta}^{(\ell')\oplus} \rangle \quad \forall \ell, \ell'$, where ℓ, ℓ' stand for $-$, 0, $+$. These scalar products can be evaluated using the orthonormal properties of the basis vectors, i.e., we get

$$\langle \vec{\eta}^{(-)\oplus} \mid \widehat{L}_- \vec{\eta}^{(-)\oplus} \rangle = 0, \quad \langle \widehat{L}_- \vec{\eta}^{(-)\oplus} \mid \vec{\eta}^{(-)\oplus} \rangle = 0.$$

$$\langle \vec{\eta}^{(-)\oplus} \mid \widehat{L}_- \vec{\eta}^{(0)\oplus} \rangle = 1, \quad \langle \widehat{L}_- \vec{\eta}^{(-)\oplus} \mid \vec{\eta}^{(0)\oplus} \rangle = 1.$$

$$\langle \vec{\eta}^{(-)\oplus} \mid \widehat{L}_- \vec{\eta}^{(+)\oplus} \rangle = 0, \langle \widehat{L}_- \vec{\eta}^{(-)\oplus} \mid \vec{\eta}^{(+)\oplus} \rangle = 0.$$

$$\langle \vec{\eta}^{(0)\oplus} \mid \widehat{L}_- \vec{\eta}^{(-)\oplus} \rangle = 1, \quad \langle \widehat{L}_- \vec{\eta}^{(0)\oplus} \mid \vec{\eta}^{(-)\oplus} \rangle = 1.$$

$$\langle \vec{\eta}^{(0)\oplus} \mid \widehat{L}_- \vec{\eta}^{(0)\oplus} \rangle = 0, \quad \langle \widehat{L}_- \vec{\eta}^{(0)\oplus} \mid \vec{\eta}^{(0)\oplus} \rangle = 0.$$

$$\langle \vec{\eta}^{(0)\oplus} \mid \widehat{L}_- \vec{\eta}^{(+)\oplus} \rangle = 0, \quad \langle \widehat{L}_- \vec{\eta}^{(0)\oplus} \mid \vec{\eta}^{(+)\oplus} \rangle = 0.$$

$$\langle \vec{\eta}^{(+)\oplus} \mid \widehat{L}_- \vec{\eta}^{(-)\oplus} \rangle = 0, \quad \langle \widehat{L}_- \vec{\eta}^{(+)\oplus} \mid \vec{\eta}^{(-)\oplus} \rangle = 0.$$

$$\langle \vec{\eta}^{(+)\oplus} \mid \widehat{L}_- \vec{\eta}^{(0)\oplus} \rangle = 0, \quad \langle \widehat{L}_- \vec{\eta}^{(+)\oplus} \mid \vec{\eta}^{(0)\oplus} \rangle = 0.$$

$$\langle \vec{\eta}^{(+)\oplus} \mid \widehat{L}_- \vec{\eta}^{(+)\oplus} \rangle = 0, \quad \langle \widehat{L}_- \vec{\eta}^{(+)\oplus} \mid \vec{\eta}^{(+)\oplus} \rangle = 0.$$

Similarly we have

$$\langle \vec{\eta}^{(-)\oplus} \mid \widehat{L}_+ \vec{\eta}^{(-)\oplus} \rangle = 0, \quad \langle \widehat{L}_+ \vec{\eta}^{(-)\oplus} \mid \vec{\eta}^{(-)\oplus} \rangle = 0.$$

$$\langle \vec{\eta}^{(-)\oplus} \mid \widehat{L}_+ \vec{\eta}^{(0)\oplus} \rangle = 0, \quad \langle \widehat{L}_+ \vec{\eta}^{(-)\oplus} \mid \vec{\eta}^{(0)\oplus} \rangle = 0.$$

$$\langle \vec{\eta}^{(-)\oplus} \mid \widehat{L}_+ \vec{\eta}^{(+)\oplus} \rangle = 0, \quad \langle \widehat{L}_+ \vec{\eta}^{(-)\oplus} \mid \vec{\eta}^{(+)\oplus} \rangle = 0.$$

$$\langle \vec{\eta}^{(0)\oplus} \mid \widehat{L}_+ \vec{\eta}^{(-)\oplus} \rangle = 0, \quad \langle \widehat{L}_+ \vec{\eta}^{(0)\oplus} \mid \vec{\eta}^{(-)\oplus} \rangle = 0.$$

$$\langle \vec{\eta}^{(0)\oplus} \mid \widehat{L}_+ \vec{\eta}^{(0)\oplus} \rangle = 0, \quad \langle \widehat{L}_+ \vec{\eta}^{(0)\oplus} \mid \vec{\eta}^{(0)\oplus} \rangle = 0.$$

$$\langle \vec{\eta}^{(0)\oplus} \mid \widehat{L}_+ \vec{\eta}^{(+)\oplus} \rangle = 1, \quad \langle \widehat{L}_+ \vec{\eta}^{(0)\oplus} \mid \vec{\eta}^{(+)\oplus} \rangle = 1.$$

$$\langle \vec{\eta}^{(+)\oplus} \mid \widehat{L}_+ \vec{\eta}^{(-)\oplus} \rangle = 0, \quad \langle \widehat{L}_+ \vec{\eta}^{(+)\oplus} \mid \vec{\eta}^{(-)\oplus} \rangle = 0.$$

$$\langle \vec{\eta}^{(+)\oplus} \mid \widehat{L}_+ \vec{\eta}^{(0)\oplus} \rangle = 1, \quad \langle \widehat{L}_+ \vec{\eta}^{(+)\oplus} \mid \vec{\eta}^{(0)\oplus} \rangle = 1.$$

$$\langle \vec{\eta}^{(+)\oplus} \mid \widehat{L}_+ \vec{\eta}^{(+)\oplus} \rangle = 0, \quad \langle \widehat{L}_+ \vec{\eta}^{(+)\oplus} \mid \vec{\eta}^{(+)\oplus} \rangle = 0.$$

These results shows that $\widehat{L}_-$ and $\widehat{L}_+$ are selfadjoint. It follows that $\widehat{L} = \widehat{L}_- + \widehat{L}_+$ is selfadjoint.

These operator is not decomposable. The subspaces $\vec{\mathcal{S}}^{(-)\oplus}, \vec{\mathcal{S}}^{(0)\oplus}$ and $\vec{\mathcal{S}}^{(+0)\oplus}$ are not invariant under any of these operators, e.g.,[4]

$$\widehat{L}_-\vec{\eta}^{(-)\oplus} = \vec{\eta}^{(0)\oplus} \notin \vec{\mathcal{S}}^{(-)\oplus},$$
$$\widehat{L}_+\vec{\eta}^{(+)\oplus} = \vec{\eta}^{(0)\oplus} \notin \vec{\mathcal{S}}^{(+)\oplus},$$
$$(\widehat{L}_- + \widehat{L}_+)\vec{\eta}^{(0)\oplus} = \vec{\eta}^{(-)\oplus} + \vec{\eta}^{(+)\oplus} \notin \vec{\mathcal{S}}^{(0)\oplus}.$$

Q24(2) Show that the tensor products of two projectors, $\widehat{P}^{(1)}$ in Hilbert space $\vec{\mathcal{H}}^{(1)}$ and $\widehat{P}^{(2)}$ in Hilbert space $\vec{\mathcal{H}}^{(2)}$, is a projector in the tensor product space $\vec{\mathcal{H}}^{\otimes} = \vec{\mathcal{H}}^{(1)} \otimes \vec{\mathcal{H}}^{(2)}$.

SQ24(2) On account of Eqs. (24.52) and (24.53) the tensor product $\widehat{P}_1 \otimes \widehat{P}_2$ is idempotent and selfadjoint in the tensor product space $\vec{\mathcal{H}}^{\otimes}$. It is therefore a projector on $\vec{\mathcal{H}}^{\otimes}$.

Q24(3) Consider the tensor product $\vec{\mathcal{H}}^{\otimes} = \vec{\mathcal{H}} \otimes \vec{\mathcal{H}}$. Let $\{\vec{\varphi}_m\}$ be an orthonormal basis for $\vec{\mathcal{H}}$. Then $\{\vec{\varphi}_m \otimes \vec{\varphi}_n\}$ is an orthonormal basis for $\vec{\mathcal{H}}^{\otimes}$. Define the **permutation operator** $\widehat{U}_p$ on $\vec{\mathcal{H}}^{\otimes}$ by[5]

$$\widehat{U}_p\Big(\sum_{m,n} c_{mn}\vec{\varphi}_m \otimes \vec{\varphi}_n\Big) := \sum_{m,n} c_{mn}\, \vec{\varphi}_n \otimes \vec{\varphi}_m.$$

(a) Show that this is a bounded operator with $\vec{\mathcal{H}}^{\otimes}$ as its domain.

(b) Show that the square of $\widehat{U}_p$ is equal to the identity, and $\widehat{U}_p$ is unitary and selfadjoint, i.e.,

$$\widehat{U}_p^2 = \widehat{I\!I}, \quad \widehat{U}_p^{\dagger} = \widehat{U}_p^{-1} = \widehat{U}_p.$$

(c) For two bounded operators $\widehat{A}$ and $\widehat{B}$ on $\vec{\mathcal{H}}^{\otimes}$ show that[6]

$$\widehat{U}_p\,(\widehat{A} \otimes \widehat{B})\,\widehat{U}_p^{\dagger} = \widehat{B} \otimes \widehat{A}.$$

(d) Show that

$$\widehat{P}^{(s)} := \frac{1}{2}\Big(\widehat{I\!I} + \widehat{U}_p\Big), \quad \widehat{P}^{(a)} := \frac{1}{2}\Big(\widehat{I\!I} - \widehat{U}_p\Big),$$

are projectors which are orthogonal to each other. Find examples of vectors in $\vec{\mathcal{H}}^{\otimes}$ which are unchanged by each of these two projectors.

---

[4] *See* the notation in Eq. (24.21).
[5] *See* §33.3 for physical applications of these operators.
[6] *See* Eq. (33.15) in §33.3.

(e) Find the eigenvalues and eigenvectors of $\widehat{U}_p$.

SQ24(3)(a) An arbitrary vector $\vec{\Phi}^{\otimes}$ in $\vec{\mathcal{H}}^{\otimes}$ can be written as

$$\vec{\Phi}^{\otimes} = \sum_{m,n} c_{mn}\vec{\varphi}_m \otimes \vec{\varphi}_n.$$

Then

$$\langle \sum_{m,n} c_{mn}\vec{\varphi}_m \otimes \vec{\varphi}_n \mid \sum_{m',n'} c_{m'n'}\vec{\varphi}_{m'} \otimes \vec{\varphi}_{n'} \rangle$$
$$= \sum_{m,n} |c_{mn}|^2 = ||\vec{\Phi}^{\otimes}||^2 < \infty.$$

Since

$$\widehat{U}_p\Big(\sum_{m,n} c_{mn}\vec{\varphi}_m \otimes \vec{\varphi}_n\Big) = \Big(\sum_{m,n} c_{mn}\,\vec{\varphi}_n \otimes \vec{\varphi}_m\Big),$$

we have

$$||\widehat{U}_p\Big(\sum_{m,n} c_{mn}\vec{\varphi}_m \otimes \vec{\varphi}_n\Big)||^2$$
$$= \langle \sum_{m,n} c_{mn}\vec{\varphi}_n \otimes \vec{\varphi}_m \mid \sum_{m',n'} c_{m'n'}\vec{\varphi}_{n'} \otimes \vec{\varphi}_{m'} \rangle$$
$$= \sum_{m,n} |c_{mn}|^2 = ||\vec{\Phi}^{\otimes}||^2.$$

Hence the operator $\widehat{U}_p$ preserves the norm and it is defined on every vector in $\vec{\mathcal{H}}^{\otimes}$. Hence it is bounded operator of the norm $||\widehat{U}_p|| = 1$.

SQ24(3)(b) Given an arbitrary vector $\vec{\Phi}^{\oplus} = \sum_{m,n} c_{mn}\vec{\varphi}_m \otimes \vec{\varphi}_n \in \vec{\mathcal{H}}^{\otimes}$ we can establish the following results:

(1) The square of $\widehat{U}_p$ is equal to itself, since

$$\widehat{U}_p^2\Big(\sum_{m,n} c_{mn}\vec{\varphi}_m \otimes \vec{\varphi}_n\Big)$$
$$= \widehat{U}_p\left(\widehat{U}_p\Big(\sum_{m,n} c_{mn}\vec{\varphi}_m \otimes \vec{\varphi}_n\Big)\right)$$
$$= \widehat{U}_p\Big(\sum_{m,n} c_{mn}\,\vec{\varphi}_n \otimes \vec{\varphi}_m\Big)$$
$$= \sum_{m,n} c_{mn}\vec{\varphi}_m \otimes \vec{\varphi}_n \;\Rightarrow\; \widehat{U}_p^2 = \widehat{I\!I}.$$

(2) From SQ24(a) we know that $\widehat{U}_p$ preserves the norm of vectors in $\vec{\mathcal{H}}^{\otimes}$. It is invertible by Theorem 17.5(1), since

$$\widehat{U}_p \vec{\Phi}^{\otimes} = \vec{0}^{\otimes} \quad \Rightarrow \quad ||\widehat{U}_p \vec{\Phi}^{\otimes}|| = ||\vec{\Phi}^{\otimes}|| = 0$$
$$\Rightarrow \vec{\Phi}^{\otimes} = \vec{0}^{\otimes}.$$

Its inverse is equal to itself, i.e., $\widehat{U}_p^{-1} = \widehat{U}_p$, since $\widehat{U}_p^2 = \widehat{I\!I}$.

(3) The adjoint of the operator is equal to its inverse since $\widehat{U}_p^{\dagger}\widehat{U}_p = \widehat{U}_p\widehat{U}_p^{\dagger} = \widehat{I\!I}$. To prove this result we have, for all $\vec{\Phi}^{\otimes}$,

$$\langle \vec{\Phi}^{\otimes} \mid \vec{\Phi}^{\otimes} \rangle = \langle \vec{\Phi}^{\otimes} \mid \widehat{U}_p\widehat{U}_p \vec{\Phi}^{\otimes} \rangle = \langle \widehat{U}_p^{\dagger} \vec{\Phi}^{\otimes} \mid \widehat{U}_p \vec{\Phi}^{\otimes} \rangle$$
$$= \langle \widehat{U}_p\widehat{U}_p^{\dagger} \vec{\Phi}^{\otimes} \mid \vec{\Phi}^{\otimes} \rangle$$
$$\Rightarrow \widehat{U}_p\widehat{U}_p^{\dagger} = \widehat{I\!I} \text{ by Eq. (18.2).}$$
$$\langle \vec{\Phi}^{\otimes} \mid \vec{\Phi}^{\otimes} \rangle = \langle \widehat{U}_p \vec{\Phi}^{\otimes} \mid \widehat{U}_p \vec{\Phi}^{\otimes} \rangle = \langle \widehat{U}_p^{\dagger}\widehat{U}_p \vec{\Phi}^{\otimes} \mid \vec{\Phi}^{\otimes} \rangle$$
$$\Rightarrow \widehat{U}_p^{\dagger}\widehat{U}_p = \widehat{I\!I}.$$

(4) Since the operator has been shown to be bounded, preserve the norm of all vectors, invertible with its inverse equal its adjoint it is therefore unitary by Theorem 18.3(1).

(5) For selfadjointness let $\vec{\Phi}'^{\otimes} = \sum_{m',n'} c'_{m'n'}\, \vec{\varphi}_{m'} \otimes \vec{\varphi}_{n'}$. Then:

$$\langle \vec{\Phi}^{\otimes} \mid \widehat{U}_p \vec{\Phi}'^{\otimes} \rangle = \langle \sum_{m,n} c_{mn} \vec{\varphi}_m \otimes \vec{\varphi}_n \mid \sum_{m',n'} c'_{m'n'} \vec{\varphi}_{n'} \otimes \vec{\varphi}_{m'} \rangle$$
$$= \sum_{m,n,m',n'} c^*_{mn} c'_{m'n'}\, \delta_{mn'}\delta_{nm'}.$$
$$\langle \widehat{U}_p \vec{\Phi}^{\otimes} \mid \vec{\Phi}'^{\otimes} \rangle = \langle \sum_{m,n} c_{mn} \vec{\varphi}_n \otimes \vec{\varphi}_m \mid \sum_{m',n'} c'_{m'n'} \vec{\varphi}_{m'} \otimes \vec{\varphi}_{n'} \rangle$$
$$= \sum_{m,n,m',n'} c^*_{mn} c'_{m'n'}\, \delta_{nm'}\delta_{mn'}.$$

It follows that $\langle \vec{\Phi}^{\otimes} \mid \widehat{U}_p \vec{\Phi}'^{\otimes} \rangle = \langle \widehat{U}_p \vec{\Phi}^{\otimes} \mid \vec{\Phi}'^{\otimes} \rangle$ which implies the selfadjointness of $\widehat{U}_p$.

From $\widehat{U}_p = \widehat{U}_p^{-1}$ we get $\widehat{U}_p = \widehat{U}_p^{\dagger} = \widehat{U}_p^{-1}$. All this confirms the previous conclusion that $\widehat{U}_p$ is unitary by Theorem 18.3(1).

SQ24(3)(c) Given an arbitrary vector $\vec{\Phi}^{\otimes} = \sum_{m,n} c_{mn}\, \vec{\varphi}_m \otimes \vec{\varphi}_n$ in $\vec{\mathcal{H}}^{\otimes}$ we have

$$\begin{aligned}
\left(\widehat{U}_p(\widehat{A} \otimes \widehat{B})\widehat{U}_p^{\dagger}\right)\vec{\Phi}^{\otimes} &= \widehat{U}_p(\widehat{A} \otimes \widehat{B})\Big(\sum_{m,n} c_{mn}\, \vec{\varphi}_n \otimes \vec{\varphi}_m\Big) \\
&= \widehat{U}_p\Big(\sum_{m,n} c_{mn}\left(\widehat{A}\vec{\varphi}_n \otimes \widehat{B}\vec{\varphi}_m\right)\Big) \\
&= \sum_{m,n} c_{mn}\left(\widehat{B}\vec{\varphi}_m \otimes \widehat{A}\vec{\varphi}_n\right) \\
&= (\widehat{B} \otimes \widehat{A})\Big(\sum_{m,n} c_{mn}\, \vec{\varphi}_m \otimes \vec{\varphi}_n\Big).
\end{aligned}$$

SQ24(3)(d) The operators $\widehat{P}^{(s)}$ and $\widehat{P}^{(a)}$ are clearly selfadjoint on account of Eq. (17.96) and the selfadjointness of $\widehat{U}_p$. Using the property $\widehat{U}_p^2 = \widehat{I\!I}$ we have

$$\begin{aligned}
\left(\widehat{P}^{(s)}\right)^2 &= \frac{1}{4}\left(\widehat{I\!I}^2 + 2\widehat{I\!I}\widehat{U}_p + \widehat{U}_p^2\right) = \widehat{P}^{(s)}. \\
\left(\widehat{P}^{(a)}\right)^2 &= \frac{1}{4}\left(\widehat{I\!I}^2 - 2\widehat{I\!I}\widehat{U}_p + \widehat{U}_p^2\right) = \widehat{P}^{(a)}.
\end{aligned}$$

Hence $\widehat{P}^{(s)}$ and $\widehat{P}^{(a)}$ are projector. They are orthogonal since

$$\widehat{P}^{(s)}\widehat{P}^{(a)} = \frac{1}{4}\left(\widehat{I\!I}^2 - \widehat{U}_p + \widehat{U}_p - \widehat{U}_p^2\right) = \widehat{0}.$$

The followings are examples of vectors unchanged by these projectors:

(1) Vectors of the form

$$\vec{\Phi}^{\otimes s} = \sum_{m,n} c_{mn}\left(\vec{\varphi}_m \otimes \vec{\varphi}_n + \vec{\varphi}_n \otimes \vec{\varphi}_m\right)$$

are unchanged by $\widehat{P}^{(s)}$ since

$$\begin{aligned}
\widehat{P}^{(s)}\vec{\Phi}^{\otimes s} &= \frac{1}{2}\Big(\sum_{m,n} c_{mn}\left(\vec{\varphi}_m \otimes \vec{\varphi}_n + \vec{\varphi}_n \otimes \vec{\varphi}_m\right) \\
&\quad + \sum_{m,n} c_{mn}\left(\vec{\varphi}_n \otimes \vec{\varphi}_m + \vec{\varphi}_m \otimes \vec{\varphi}_n\right)\Big) \\
&= \sum_{m,n} c_{mn}\left(\vec{\varphi}_m \otimes \vec{\varphi}_n + \vec{\varphi}_n \otimes \vec{\varphi}_m\right) = \vec{\Phi}^{\otimes s}.
\end{aligned}$$

These vectors are shown in Eq. (33.19) and they are said to be symmetrical in Definition 33.3.1(1).

(2) Vectors of the form

$$\vec{\Phi}^{\otimes a} = \sum_{m,n} c_{mn} \left(\vec{\varphi}_m \otimes \vec{\varphi}_n - \vec{\varphi}_n \otimes \vec{\varphi}_m \right)$$

are similarly unchanged by $\widehat{P}^{(a)}$. These vectors are shown in Eq. (33.20) and they are said to be antsymmetrical in Definition 33.3.1(1).

SQ24(3)(e) Consider the eigenvalue equation of $\widehat{U}_p$, i.e., $\widehat{U}_p \vec{\Phi}^{\otimes}_{\lambda} = \lambda \vec{\Phi}^{\otimes}_{\lambda}$, where $\lambda \in I\!R$. Then $\widehat{U}_p^2 \vec{\Phi}^{\otimes}_{\lambda} = \lambda^2 \vec{\Phi}^{\otimes}_{\lambda}$. Since $\widehat{U}_p^2 = \widehat{I\!I}$ and the eigenvalues of $\widehat{U}_p$ are real (since $\widehat{U}_p$ is selfadjoint) we can immediately conclude that $\lambda^2 = 1$ which implies that λ is equal to either 1 or -1. Their corresponding eigenvectors are $\widehat{P}^{(s)} \vec{\Phi}^{\otimes}$ and $\widehat{P}^{(a)} \vec{\Phi}^{\otimes}$, i.e.,

$$\widehat{U}_p \big(\widehat{P}^{(s)} \vec{\Phi}^{\otimes} \big) = + \big(\widehat{P}^{(s)} \vec{\Phi}^{\otimes} \big),$$
$$\widehat{U}_p \big(\widehat{P}^{(a)} \vec{\Phi}^{\otimes} \big) = - \big(\widehat{P}^{(a)} \vec{\Phi}^{\otimes} \big).$$

Chapter 25

Pure States

Q25(1) Explain the concept of pure states.

SQ25(1) Quantum systems satisfy property QMP5.3(1), i.e., not all observables are simultaneously measurable. Simultaneously measurable observables are said to be compatible. A maximum amount of information about the system corresponds to a set of simultaneously measured values of a maximum set of compatible discrete observables of the system, known as a complete set of discrete observables. Such an amount of information characterises a state of the system. States characterised by such a maximum amount of information about the system are called **pure states**. If we do not have a maximum amount of information about the system we cannot determine a pure state. Such a situation does occur in many practical cases. It is still desirable to have a characterisation of the system based on the information practically available. Such a characterisation which is based on less than a maximum set of data about the system leads to the notation of **mixed states** which is discussed in Chapter 31.

Note that a measurement of a continuous observables like position and linear momentum cannot produce a precise value (*see* P4.3.2(5) and C28.2(3)). Hence they are not used in the above discussion of the concept of pure states.

Quantum Mechanics: Problems and Solutions
K. Kong Wan

ISBN 978-981-4800-72-3 (Paperback), 978-0-429-29647-5 (eBook)
www.jennystanford.com

Q25(2) Explain why only unit vectors are used for state description in Postulate 25.1(PS).[1]

SQ25(2) Theorem 22.1(1) tells us that a unit vector in a Hilbert space together with the spectral function of a selfadjoint operator in the Hilbert space can generate a probability distribution function. It follows that if we describe a state by a unit vector and an observable by a selfadjoint operator we can generate a probability distribution function which can be taken to describe the probability distribution of the values of the observable. Such a description of states and observables is demonstrated in the model theory for electron spin in §14.1.1. A formal statement that observables correspond to the selfadjoint operators is given by Postulate 26.1(OV).

A vector which is not normalised cannot be used in Theorem 22.1(1) to produce a probability distribution function. It needs to be normalised first.

Q25(3) Explain why pure states do not correspond one-to-one to unit vectors in the state space.

SQ25(3) According to Theorem 22.1(1) two unit vectors which differ only by a *phase factor*, i.e., a multiplicative constant of magnitude 1, would generate the same probability distribution function. It is therefore not possible to distinguish two such unit vectors physically as far as state description is concerned. In other words all the unit vectors different by a phase factor would describe the same state. It follows that pure states do not correspond one-to-one to unit vectors.

All the unit vectors which differ only by a phase factor lie in a one-dimensional subspace. Hence pure states correspond to one-dimensional subspaces. Since one-dimensional subspaces correspond to one-dimensional projectors we have a correspondence between pure states and one-dimensional projectors. The correspondence is one-to-one for orthodox quantum system defined by Definition 26.1(1).

[1] *See* also the discussion in §14.1.1 and §22.1.

Chapter 26

Observables and Their Values

Q26(1) Discuss the fundamental differences between classical observables and quantum observables.

SQ26(1)

Classical observables possess the following properties:

(1) They are all compatible, i.e., they are all simultaneously measurable.

(2) Classical observables are related to the state in that they are described by numerical functions defined on the classical state space, i.e., they are numerical functions of the state. It follows that a given state would determine the values of all observables. In other words a classical system in a given state would possess a value of every observables. These values can be revealed by measurement.

(3) As the the state evolves in time observables would evolve in time accordingly. The values of observables at a later time are determined by their initial values.

Quantum observables possess the following contrasting properties:

(1) They are not all compatible, i.e., they are not all simultaneously measurable. Only a limited number of quantum observables are compatible.

Quantum Mechanics: Problems and Solutions
K. Kong Wan

ISBN 978-981-4800-72-3 (Paperback), 978-0-429-29647-5 (eBook)
www.jennystanford.com

(2) Observables are not mathematically related to the state in such a way that their values are determined by the state, e.g., they are not functions of the state. A state does not determine the value of an arbitrary observable. A state can only generate a probability distribution of the values of an observable. The prescription for generating such a probability distribution needs to be explicitly stated. Details are discussed in Chapter 28.

(3) Time evolution of the state does not automatically determine the time evolution of observables. The future values of the observables are not generally determined by their initial values. Details are discussed in Chapter 29.

The different relationship between states and observables in classical and quantum mechanics entails a fundamental difference in the meaning of conservation laws for classical and quantum observables. This will become clear after the discussion of quantum time evolution in Chapter 29.[1]

Q26(2) What are the measurable values of a function $f(A)$ of a discrete observable A described by a selfadjoint operator $\widehat{A}$ which has a discrete spectrum $sp_d = \{a_1, a_2, \ldots\}$?

SQ26(2) Given an observable A described by a selfadjoint operator $\widehat{A}$ the observable corresponding to the function $f(A)$ of A is described by the operator $f(\widehat{A})$. This is stated in C26.1(6). When $\widehat{A}$ has a purely discrete spectrum $\{a_m\}$ the operator $f(\widehat{A})$ has the following spectral decomposition[2]:

$$f(\widehat{A}) = \sum_m f(a_m)\, \widehat{P}^{\widehat{A}}(a_m).$$

It follows that $f(\widehat{A})$ has a discrete set of eigenvalues $f(a_m)$ which are then the measurable values of the observable $f(A)$.

Q26(3) Give a brief account of the concept of propositions in quantum mechanics.

---

[1] *See* Definition 29.1.2(1) and Q29(6) in Exercises and Problems for Chapter 29.

[2] *See* §20.5 for discussion on functions of selfadjoint operators. Definition 13.3.3(1), i.e., Eqs. (13.37) and (13.38), applies to discrete observables.

SQ26(3) Physically a proposition is a statement about the system which is either true or false.[3] An experiment can be performed to test whether a proposition is true or false. Such an experiment is called a yes-no experiment. In quantum mechanics a proposition is a discrete observable described by a projector. A proposition has only two values, i.e., 1 and 0 which correspond to the eigenvalues of its associated projector. The yes-no experiment is arranged such that the value 1 corresponds to the yes answer and the value 0 corresponds to the no answer to the proposition.

A general observable has a set of propositions associated with it. These propositions correspond to the spectral projectors of the selfadjoint operator representing the observable.

Q26(4) The spectral projector $\widehat{M}^{\hat{x}}(\Lambda)$ of the position operator $\widehat{x}$ for an interval Λ is defined by a characteristic function in Eq. (20.28). What is the physical meaning of $\widehat{M}^{\hat{x}}(\Lambda)$ as a proposition? What physical devices are capable of measuring $\widehat{M}^{\hat{x}}(\Lambda)$?

SQ26(4) The spectral projector $\widehat{M}^{\hat{x}}(\Lambda)$ describes the proposition that a measurement of the position of the particle will result in a value τ in the interval Λ. In other words $\widehat{M}^{\hat{x}}(\Lambda)$ represents the proposition that a measurement of the position of the particle will result in a value in the interval Λ. These propositions are called local position observables. Detectors such as Geiger counters can serve as a measuring device to carry out the yes-no experiment to measure the proposition. A detailed discussion of position measurement is given in §30.2.2.

Q26(5) A state ϕ^s corresponds to the one-dimensional projector $\widehat{P}_{\vec{\phi}}$ generated by the state vector $\vec{\phi}$. What is the meaning of the proposition corresponding to the projector $\widehat{P}_{\vec{\phi}}$?

SQ26(5) The projector $\widehat{P}_{\vec{\phi}}$ describes the proposition that the system is in the state ϕ^s described by the state vector $\vec{\phi}$. An ideal measurement of the proposition resulting in the value 1 would tell us that the state is described by the unit vector $\vec{\phi}$ immediately after the measurement.

[3] Isham pp. 168–178 for a discussion of various interpretations. This kind of observables also exist in classical mechanics (*see* Isham pp. 61–65).

Chapter 27

Canonical Quantisation

Q27(1) Show that Hamilton's equations of motion in Eq. (27.8) for a classical harmonic oscillator whose Hamiltonian is given by Eq. (27.11) are equivalent to Newton's equation of motion.

SQ27(1) For a classical oscillator we have $q_j = x$, $p_{cj} = p$ and $H_{ho} = p^2/2m + m\omega^2 x^2/2$. The Hamilton's equations become

$$\frac{dx}{dt} = \frac{\partial H_{ho}(x, p)}{\partial p} = \frac{1}{m}p, \quad \frac{dp}{dt} = -\frac{\partial H_{ho}(x, p)}{\partial x} = -m\omega^2 x$$

$$\Rightarrow \quad \frac{d^2x}{dt^2} = \frac{1}{m}\frac{dp}{dt} = -\omega^2 x.$$

Newton's equation of motion is

$$m\frac{dx^2}{dt^2} = -\frac{dV(x)}{dx}.$$

For the oscillator the potential energy is $V(x) = m\omega^2 x^2/2$. The above Newton's equation becomes

$$\frac{dx^2}{dt^2} = -\omega^2 x,$$

which agrees with Hamilton's equations of motion.

Q27(2) Verify the properties of Poisson brackets shown in Eqs. (27.54) to (27.59).

Quantum Mechanics: Problems and Solutions
K. Kong Wan

ISBN 978-981-4800-72-3 (Paperback), 978-0-429-29647-5 (eBook)
www.jennystanford.com

SQ27(2) It is obvious that

$$\{A, A\} = 0, \qquad \{A, c\} = 0,$$
$$\{A, B\} = -\{B, A\}, \qquad \{A, cB\} = c\{A, B\}.$$

Next we have

$$\begin{aligned}\{A, B + C\} &= \sum_{j=1}^{3}\left(\frac{\partial A}{\partial x_j}\frac{\partial (B + C)}{\partial p_{cj}} - \frac{\partial A}{\partial p_{cj}}\frac{\partial (B + C)}{\partial x_j}\right)\\ &= \sum_{j=1}^{3}\left(\frac{\partial A}{\partial x_j}\frac{\partial B}{\partial p_{cj}} - \frac{\partial A}{\partial p_{cj}}\frac{\partial B}{\partial x_j}\right)\\ &\quad + \sum_{j=1}^{3}\left(\frac{\partial A}{\partial x_j}\frac{\partial C}{\partial p_{cj}} - \frac{\partial A}{\partial p_{cj}}\frac{\partial C}{\partial x_j}\right)\\ &= \{A, B\} + \{A, C\}.\end{aligned}$$

$$\begin{aligned}\{A, BC\} &= \sum_{j=1}^{3}\left(\frac{\partial A}{\partial x_j}\frac{\partial (BC)}{\partial p_{cj}} - \frac{\partial A}{\partial p_{cj}}\frac{\partial (BC)}{\partial x_j}\right)\\ &= \sum_{j=1}^{3}\left(\frac{\partial A}{\partial x_j}\frac{\partial B}{\partial p_{cj}}C - \frac{\partial A}{\partial p_{cj}}\frac{\partial B}{\partial x_j}C\right)\\ &\quad + \sum_{j=1}^{3}\left(B\frac{\partial A}{\partial x_j}\frac{\partial C}{\partial p_{cj}} - B\frac{\partial A}{\partial p_{cj}}\frac{\partial C}{\partial x_j}\right)\\ &= \{A, B\}C + B\{A, C\}.\end{aligned}$$

The equations for $\{A + B, C\}$ and $\{AB, C\}$ are similarly proved.

Q27(3) Verify the Poisson bracket relations in Eq. (27.61) between the components of the canonical angular momentum $\vec{L}_c$.

SQ27(3) Using properties in Eqs. (27.54) to (27.59) and the Poisson brackets of the canonical variables in Eq. (27.60) we get

$$\begin{aligned}\{L_{cx}, L_{cy}\} &= \{yp_{cz} - zp_{cy}, zp_{cx} - xp_{cz}\}\\ &= \{yp_{cz}, zp_{cx} - xp_{cz}\} - \{zp_{cy}, zp_{cx} - xp_{cz}\}\\ &= \{yp_{cz}, zp_{cx}\} + \{zp_{cy}, xp_{cz}\} = -yp_{cx} + xp_{cy}\\ &= L_{cz},\end{aligned}$$

$$\begin{aligned}\{L_{cz}, L_{cx}\} &= \{xp_{cy} - yp_{cx}, yp_{cz} - zp_{cy}\} \\ &= \{xp_{cy}, yp_{cz} - zp_{cy}\} - \{yp_{cx}, yp_{cz} - zp_{cy}\} \\ &= \{xp_{cy}, yp_{cz}\} + \{yp_{cx}, zp_{cy}\} = -xp_{cz} + zp_{cx} \\ &= L_{cy},\end{aligned}$$

$$\begin{aligned}\{L_{cy}, L_{cz}\} &= \{zp_{cx} - xp_{cz}, xp_{cy} - yp_{cx}\} \\ &= \{zp_{cx}, xp_{cy} - yp_{cx}\} - \{xp_{cz}, xp_{cy} - yp_{cx}\} \\ &= \{zp_{cx}, xp_{cy}\} + \{xp_{cz}, yp_{cx}\} = -zp_{cy} + yp_{cz} \\ &= L_{cx}.\end{aligned}$$

Q27(4) Show that the equation of motion (27.53) in terms of Poisson bracket reduces to the Hamilton's equations when we replace A by x_i and p_i.

SQ27(4) Bearing in mind that x_i and p_{cj} are independent, i.e.,

$$\frac{\partial x_i}{\partial x_j} = \delta_{ij}, \quad \frac{\partial p_{ci}}{\partial p_{cj}} = \delta_{ij}, \quad \frac{\partial p_i}{\partial x_j} = 0, \quad \frac{\partial x_i}{\partial p_{cj}} = 0,$$

we get the corresponding Hamilton's equations of motion, i.e.,

$$\frac{dx_i}{dt} = \sum_{j=1}^{3} \left(\frac{\partial x_i}{\partial x_j}\frac{\partial H}{\partial p_{cj}} - \frac{\partial x_i}{\partial p_{cj}}\frac{\partial H}{\partial x_j} \right) = \frac{\partial H}{\partial p_{ci}},$$

$$\frac{dp_{ci}}{dt} = \sum_{j=1}^{3} \left(\frac{\partial p_{ci}}{\partial x_j}\frac{\partial H}{\partial p_{cj}} - \frac{\partial p_{ci}}{\partial p_{cj}}\frac{\partial H}{\partial p_{cj}} \right) = -\frac{\partial H}{\partial p_{ci}}.$$

Q27(5) Show that a classical observable is a *constant of motion*, i.e., it is time-independent, if it has a zero Poisson bracket with the Hamiltonian.[1]

SQ27(5) If A has a zero Poisson bracket with the Hamiltonian, then $dA/dt = 0$ by Eq. (27.53). The observable is time-independent, i.e., it is a constant of motion.

Q27(6) Verify Eq. (27.70). Show that Postulate 27.2(CQ) as expressed in Eq. (27.69) cannot be valid without the imaginary number i.

---

[1]Recall that we confine ourselves to observables which are not explicitly time dependent unless otherwise is stated.

SQ27(6) First let us have a quick look at how Eq. (27.70) can come about. Using the selfadjointness of $\widehat{Q}_i$ and $\widehat{P}_j$ and treating Eqs. (17.100) and (17.101) as equalities we get

$$[\widehat{Q}_i, \widehat{P}_j]^\dagger = \left(\widehat{Q}_i\widehat{P}_j\right)^\dagger - \left(\widehat{P}_j\widehat{Q}_i\right)^\dagger = \widehat{P}_j\widehat{Q}_i - \widehat{Q}_i\widehat{P}_j = -[\widehat{Q}_i, \widehat{P}_j].$$

This shows that $[\widehat{Q}_i, \widehat{P}_j]$ is not selfadjoint. Let us suppose

$$[\widehat{Q}_j, \widehat{P}_j] = \hbar\widehat{I\!I}.$$

Then the right-hand side of the above equation is selfadjoint while the left-hand side is not. Explicitly we have

$$\begin{aligned}[\widehat{Q}_j, \widehat{P}_j] = \hbar\widehat{I\!I} \quad &\Rightarrow \quad [\widehat{Q}_j, \widehat{P}_j]^\dagger = \hbar\widehat{I\!I}^\dagger = \hbar\widehat{I\!I} = [\widehat{Q}_i, \widehat{P}_j] \\ &\Rightarrow \quad -[\widehat{Q}_j, \widehat{P}_j,] = [\widehat{Q}_j, \widehat{P}_j],\end{aligned}$$

which is a contradiction since $[\widehat{Q}_j, \widehat{P}_j] \neq \widehat{0}$. It follows that the above equation cannot be true. An additional factor i on the right-hand side of the commutation relation as stated in Postulate 27.2(CQ), i.e., $[\widehat{Q}_j, \widehat{P}_j] = i\hbar\widehat{I\!I}$, would produce the desired minus sign on both sides of the commutation relation to avoid this contradiction.

Now let us examine Eq. (27.70) more carefully again. This equation is meant to hold in a restricted domain and not on the entire Hilbert space since the operators involved are unbounded. The same applies to all the equations presented above. However, the calculation above can be justified without a full specification of the domain of operations of those equations if we can find a subset of vectors on which all those equations would hold. This point is discussed in P27.2(4). To be definite let us consider the situation in $\vec{L}^2(I\!R)$ where we have $\widehat{Q} = \widehat{x}$ and $\widehat{P} = \widehat{p}$. The Schwartz space $\vec{S}_s(I\!R)$ is invariant under $\widehat{x}$ and $\widehat{p}$.[2] Restricted to such an invariant subspace Eqs. (17.100) and (17.101) become equalities which justify subsequent calculation.

Q27(7) Show that $\widehat{U}$ in Eq. (27.79) is unitary and that $\widehat{Q}', \widehat{P}'$ in Eq. (27.78) are the unitary transforms of $\widehat{Q}, \widehat{P}$ generated by this unitary operator.

---

[2]*See* §16.1.2.4 for the definition of Schwartz functions. *See* also E.17.3.2.5(2), Eq. (17.62) and P27.2(4).

SQ27(7) Since $f(x)$ is a real-valued function of $x \in I\!R$ the following integral

$$-\frac{1}{\hbar}\int^{x} f(x)dx$$

define a selfadjoint multiplication operator. The operator $\widehat{U}$ in Eq. (27.79) is an exponential function of this selfadjoint multiplication operator and is hence unitary by Eq. (21.3). We can also verify that $\widehat{U}$ in Eq. (27.79) satisfies Theorem 18.3(1), i.e., $\widehat{U}^{\dagger} = \widehat{U}^{-1}$.

For the unitary transforms we first have $\widehat{Q}' = \widehat{U}\widehat{Q}\widehat{U}^{\dagger} = \widehat{Q}$ since $\widehat{Q} := \widehat{x}$ commutes with $\widehat{U}$. Next we have, for $\vec{\phi} \in \mathcal{D}(\widehat{p})$,

$$\begin{aligned}\widehat{p}\,\widehat{U}^{\dagger}\vec{\phi} &:= -i\hbar\frac{d}{dx}\left(\exp\left\{\frac{i}{\hbar}\int^{x} f(x)dx\right\}\phi(x)\right)\\ &= -i\hbar\left(\frac{i}{\hbar}f(x)\exp\left\{\frac{i}{\hbar}\int^{x} f(x)dx\right\}\phi(x)\right.\\ &\qquad\left.+\exp\left\{\frac{i}{\hbar}\int^{x} f(x)dx\right\}\frac{d\phi(x)}{dx}\right).\end{aligned}$$

It follows that

$$\begin{aligned}&\widehat{p}\,\widehat{U}^{\dagger}\vec{\phi} = f(\widehat{x})\widehat{U}^{\dagger}\vec{\phi} + \widehat{U}^{\dagger}\widehat{p}\vec{\phi}\\ \Rightarrow\ &\widehat{p}' = \widehat{U}\,\widehat{p}\,\widehat{U}^{\dagger}\vec{\phi} = \widehat{U}\,f(\widehat{x})\widehat{U}^{\dagger}\vec{\phi} + \widehat{U}\,\widehat{U}^{\dagger}\,\widehat{p}\vec{\phi} = \big(f(\widehat{x}) + \widehat{p}\,\big)\vec{\phi}.\end{aligned}$$

We have used the fact that $\widehat{U}$ commutes with $f(\widehat{x})$ and $\widehat{U}\,\widehat{U}^{\dagger} = \widehat{I\!I}$.

Q27(8) Prove Eq. (27.104) by the method of induction.

SQ27(8) The method of induction is a method used to prove a sequence of statements to be true.[3] We can apply this method here. So, to prove

$$[\widehat{x}, \widehat{p}^{n}] = i\hbar n\widehat{p}^{n-1} \quad \text{and} \quad [\widehat{p}, \widehat{x}^{n}] = -i\hbar n\widehat{x}^{n-1}$$

we proceed in three steps:

Step 1: The above equations are true for $n = 1$, on account of the canonical quantisation rule, i.e.,

$$[\widehat{x}, \widehat{p}\,] = i\hbar, \quad [\,\widehat{p}, \widehat{x}\,] = -i\hbar.$$

Step 2: Suppose the equations are true for any n. We desire to prove that they are also true for $n + 1$, i.e.,

$$\begin{aligned}[\widehat{x}, \widehat{p}^{\,n}\,] = i\hbar n\widehat{p}^{n-1} &\Rightarrow [\widehat{x}, \widehat{p}^{(n+1)}\,] = i\hbar\,(n+1)\widehat{p}^{\,n},\\ [\,\widehat{p}, \widehat{x}^{n}] = -i\hbar\,n\widehat{x}^{n-1} &\Rightarrow [\,\widehat{p}, \widehat{x}^{(n+1)}] = -i\hbar\,(n+1)\widehat{x}^{\,n}.\end{aligned}$$

[3]Such a method is set out in SQ17(7) and used also in SQ27(13).

All we need is to reduce $[\widehat{x}, \widehat{p}^{(n+1)}]$, $[\widehat{p}, \widehat{x}^{(n+1)}]$ in terms of $[\widehat{x}, \widehat{p}^n]$ and $[\widehat{p}, \widehat{x}^n]$ using the formula

$$[\widehat{A}, \widehat{B}\widehat{C}] = [\widehat{A}, \widehat{B}]\widehat{C} + \widehat{B}[\widehat{A}, \widehat{C}].$$

We have

$$\begin{aligned}[\widehat{x}, \widehat{p}^{(n+1)}] &= [\widehat{x}, \widehat{p}^n\, \widehat{p}] = [\widehat{x}, \widehat{p}^n]\,\widehat{p} + \widehat{p}^n\,[\widehat{x}, \widehat{p}] \\ &= \left(i\hbar n\widehat{p}^{(n-1)}\right)\,\widehat{p} + (\widehat{p}^n)\, i\hbar \\ &= i\hbar\,(n+1)\widehat{p}^n,\end{aligned}$$

$$\begin{aligned}[\widehat{p}, \widehat{x}^{(n+1)}] &= [\widehat{p}, \widehat{x}^n]\widehat{x} + \widehat{x}^n\,[\widehat{p}, \widehat{x}] \\ &= \left(-i\hbar n\widehat{x}^{(n-1)}\right)\,\widehat{x} + \widehat{x}^n\,(-i\hbar) \\ &= -i\hbar\,(n+1)\widehat{x}^n.\end{aligned}$$

Step 3: We can now conclude that the desired equations are true for all n.

Q27(9) Verify the commutation relations in Eqs. (27.106) and (27.107).

SQ27(9) Consider an observable $\widehat{A}$ which is a polynomial in $\widehat{x}$ and $\widehat{p}$, i.e.,

$$\widehat{A} = \sum_{n=1}^{N}\sum_{m=1}^{M} c_{mn}\,\widehat{x}^n\,\widehat{p}^m.$$

Using Eq. (27.104) we get

$$\begin{aligned}[\widehat{x}, \widehat{A}] &= \sum_{m=1}^{M} c_{mn}[\widehat{x}, \widehat{x}^n\,\widehat{p}^m] = \sum_{m=1}^{M} c_{mn}\,\widehat{x}^n\,[\widehat{x}, \widehat{p}^m] \\ &= \sum_{m=1}^{M} c_{mn}\,\widehat{x}^n\,\left(i\hbar\, m\,\widehat{p}^{(m-1)}\right) = i\hbar\,\frac{\partial\widehat{A}}{\partial\widehat{p}}.\end{aligned}$$

$$\begin{aligned}[\widehat{p}, \widehat{A}] &= \sum_{n=1}^{N}\sum_{m=1}^{M} c_{mn}\,[\widehat{p}, \widehat{x}^n\,\widehat{p}^m] = \sum_{n=1}^{N}\sum_{m=1}^{M} c_{mn}\,[\widehat{p}, \widehat{x}^n]\,\widehat{p}^m \\ &= \sum_{m=1}^{M} c_{mn}\,\left(i\hbar\, n\,\widehat{x}^{(n-1)}\right)\,\widehat{p}^m = -i\hbar\,\frac{\partial\widehat{A}}{\partial\widehat{x}}.\end{aligned}$$

Note that the differentiations are just formal expressions.

Q27(10) Verify the commutation relations in Eqs. (27.108), (27.109) and (27.110).

SQ27(10) Treating $\widehat{H}$ as a polynomial functions of $\widehat{x}$ and $\widehat{p}$ and using Eqs. (27.106) and (27.107) we get

$$[\,\widehat{x},\,\widehat{H}\,] = i\hbar\,\frac{\partial\,\widehat{H}}{\partial\,\widehat{p}}. \quad \text{and} \quad [\,\widehat{p},\,\widehat{H}\,] = -i\hbar\,\frac{\partial\,\widehat{H}}{\partial\,\widehat{x}}.$$

For the harmonic oscillator Hamiltonian we get

$$[\,\widehat{x},\,\widehat{H}_{ho}\,] = i\hbar\,\frac{\partial\,\widehat{H}_{ho}}{\partial\,\widehat{p}} = \frac{i\hbar}{m}\,\widehat{p},$$

$$[\,\widehat{p},\,\widehat{H}_{ho}\,] = -i\hbar\,\frac{\partial\,\widehat{H}_{ho}}{\partial\,\widehat{x}} = -i\hbar m\omega^2\,\widehat{x}.$$

Q27(11) Verify the commutation relations in Eqs. (27.111) to (27.114) for angular momentum operators.

SQ27(11) From Eqs. (27.64) to (27.68) and the canonical commutation relations for the canonical variables we get[4]

$$\begin{aligned}
[\widehat{L}_x,\,\widehat{L}_y\,] &= [\,\widehat{y}\,\widehat{p}_z - \widehat{z}\,\widehat{p}_y,\,\widehat{z}\,\widehat{p}_x - \widehat{x}\,\widehat{p}_z\,] \\
&= [\,\widehat{y}\,\widehat{p}_z,\,\widehat{z}\,\widehat{p}_x - \widehat{x}\,\widehat{p}_z\,] - [\,\widehat{z}\,\widehat{p}_y,\,\widehat{z}\,\widehat{p}_x - \widehat{x}\,\widehat{p}_z\,] \\
&= [\,\widehat{y}\,\widehat{p}_z,\,\widehat{z}\,\widehat{p}_x\,] + [\,\widehat{z}\,\widehat{p}_y,\,\widehat{x}\,\widehat{p}_z\,] \\
&= -i\hbar\widehat{y}\,\widehat{p}_x + i\hbar\widehat{x}\,\widehat{p}_y = i\hbar\widehat{L}_z. \\
[\widehat{L}_z,\,\widehat{L}_x\,] &= [\,\widehat{x}\,\widehat{p}_y - \widehat{y}\,\widehat{p}_x,\,\widehat{y}\,\widehat{p}_z - \widehat{z}\,\widehat{p}_y\,] \\
&= [\,\widehat{x}\,\widehat{p}_y,\,\widehat{y}\,\widehat{p}_z - \widehat{z}\,\widehat{p}_y\,] - [\,\widehat{y}\,\widehat{p}_x,\,\widehat{y}\,\widehat{p}_z - \widehat{z}\,\widehat{p}_y\,] \\
&= [\,\widehat{x}\,\widehat{p}_y,\,\widehat{y}\,\widehat{p}_z\,] + [\,\widehat{y}\,\widehat{p}_x,\,\widehat{z}\,\widehat{p}_y\,] \\
&= -i\hbar\widehat{x}\,\widehat{p}_z + i\hbar\widehat{z}\,\widehat{p}_x = i\hbar\widehat{L}_y. \\
[\widehat{L}_y,\,\widehat{L}_z\,] &= [\,\widehat{z}\,\widehat{p}_x - \widehat{x}\,\widehat{p}_z,\,\widehat{x}\,\widehat{p}_y - y\,\widehat{p}_x\,] \\
&= [\,\widehat{z}\,\widehat{p}_x,\,\widehat{x}\,\widehat{p}_y - y\,\widehat{p}_x\,] - [\,\widehat{x}\,\widehat{p}_z,\,\widehat{x}\,\widehat{p}_y - y\,\widehat{p}_x\,] \\
&= [\,\widehat{z}\,\widehat{p}_x,\,\widehat{x}\,\widehat{p}_y\,] + [\,\widehat{x}\,\widehat{p}_z,\,\widehat{y}\,\widehat{p}_x\,] \\
&= -i\hbar\,\widehat{z}\,\widehat{p}_y + i\hbar\,\widehat{y}\,\widehat{p}_z = i\hbar\widehat{L}_x.
\end{aligned}$$

[4] *See* SQ27(3).

$$\begin{aligned}[\widehat{L}_z, \widehat{L}^2] &= [\widehat{L}_z, \widehat{L}_x^2] + [\widehat{L}_z, \widehat{L}_y^2] + [\widehat{L}_z, \widehat{L}_z^2] \\ &= [\widehat{L}_z, \widehat{L}_x^2] + [\widehat{L}_z, \widehat{L}_y^2] \\ &= \widehat{L}_x[\widehat{L}_z, \widehat{L}_x] + [\widehat{L}_z, \widehat{L}_x]\widehat{L}_x \\ &\quad \widehat{L}_y[\widehat{L}_z, \widehat{L}_y] + [\widehat{L}_z, \widehat{L}_y]\widehat{L}_y \\ &= i\hbar(\widehat{L}_x\widehat{L}_y + \widehat{L}_y\widehat{L}_x) - i\hbar(\widehat{L}_y\widehat{L}_x + \widehat{L}_x\widehat{L}_y) \\ &= \widehat{0}.\end{aligned}$$

Q27(12) Verify the commutation relations in Eq. (27.122).

SQ27(12) Using Eqs. (27.64) to (27.68) and (27.121) we get

$$[\,\widehat{a}, \widehat{N}\,] = [\,\widehat{a}, \widehat{a}^\dagger \widehat{a}\,] = [\,\widehat{a}, \widehat{a}^\dagger\,]\widehat{a} = \widehat{a}.$$

$$[\,\widehat{a}^\dagger, \widehat{N}\,] = [\,\widehat{a}^\dagger, \widehat{a}^\dagger \widehat{a}\,] = \widehat{a}^\dagger[\,\widehat{a}^\dagger, \widehat{a}\,] = -\widehat{a}^\dagger.$$

Q27(13) Prove, by the method of induction, that the eigenvectors in Eq. (27.126) are normalised.

SQ27(13) The eigenvector $\vec{\varphi}_0$ is taken as normalised. Then: (1) $\vec{\varphi}_1$ is normalised since

$$\begin{aligned}\langle \vec{\varphi}_1 \mid \vec{\varphi}_1 \rangle &= \langle \widehat{a}^\dagger \vec{\varphi}_0 \mid \widehat{a}^\dagger \vec{\varphi}_0 \rangle = \langle \vec{\varphi}_0 \mid \widehat{a}\widehat{a}^\dagger \vec{\varphi}_0 \rangle \\ &= \langle \vec{\varphi}_0 \mid (\widehat{a}^\dagger \widehat{a} + \widehat{I\!I})\vec{\varphi}_0 \rangle = \langle \vec{\varphi}_0 \mid \widehat{I\!I} \vec{\varphi}_0 \rangle = 1.\end{aligned}$$

(2) Assume that $\vec{\varphi}_n = (\widehat{a}^\dagger)^n \vec{\varphi}_0 / \sqrt{n!}$ is normalised, i.e., $\langle \vec{\varphi}_n \mid \vec{\varphi}_n \rangle = 1$. Then we have

$$\begin{aligned}\vec{\varphi}_{n+1} &= \frac{1}{\sqrt{(n+1)!}} (\widehat{a}^\dagger)^{n+1} \vec{\varphi}_0 \\ &= \frac{1}{\sqrt{n+1}} \widehat{a}^\dagger \left(\frac{1}{\sqrt{n!}} (\widehat{a}^\dagger)^n \vec{\varphi}_0 \right) = \frac{1}{\sqrt{n+1}} \widehat{a}^\dagger \vec{\varphi}_n \\ \Rightarrow \quad \langle \vec{\varphi}_{n+1} \mid \vec{\varphi}_{n+1} \rangle &= \frac{1}{n+1} \langle \widehat{a}^\dagger \vec{\varphi}_n \mid \widehat{a}^\dagger \vec{\varphi}_n \rangle = \frac{1}{n+1} \langle \vec{\varphi}_n \mid \widehat{a}\widehat{a}^\dagger \vec{\varphi}_n \rangle \\ &= \frac{1}{n+1} \langle \vec{\varphi}_n \mid (\widehat{a}^\dagger \widehat{a} + \widehat{I\!I})\vec{\varphi}_n \rangle \\ &= \frac{1}{n+1} \langle \vec{\varphi}_n \mid (n+1)\vec{\varphi}_n \rangle = 1.\end{aligned}$$

We can conclude that all $\vec{\varphi}_n$ are normalised by the method of induction.

Q27(14) Let $\vec{\Phi}$ be a vector in $\vec{L}^2(I\!R^3)$ defined by the product of a function of the radial variable r and a spherical harmonics $Y_{\ell,m_\ell}(\theta,\varphi)$, i.e.,

$$\vec{\Phi} := \Phi(r,\theta,\varphi) = \phi(r)Y_{\ell,m_\ell}(\theta,\varphi).$$

Let $\vec{\Psi}$ be another vector defined in the same way, i.e.,

$$\vec{\Psi} := \Psi(r,\theta,\varphi) = \psi(r)Y_{\ell,m_\ell}(\theta,\varphi).$$

Furthermore the functions $\phi(r)$ and $\psi(r)$ satisfy the boundary condition

$$\lim_{r\to 0} r|\phi(r)| = 0 \quad \text{and} \quad \lim_{r\to 0} r|\psi(r)| = 0. \tag{$*$}$$

Working in spherical coordinates show that[5]

$$\langle \vec{\Psi} \mid \widehat{p}_r\vec{\Phi}\,\rangle = \langle \widehat{p}_r\vec{\Psi} \mid \vec{\Phi}\,\rangle,$$

where $\widehat{p}_r$ is the radial momentum operator introduced by Eq. (27.148).[6] Explain why $\widehat{p}_r$ can be symmetric but not selfadjoint.

SQ27(14) For functions $\Psi(r,\theta,\varphi)$ and $\Phi(r,\theta,\varphi)$ in $L^2(I\!R^3)$ the scalar product integral in spherical coordinates r,θ,φ is with respect to the volume element $dx^3 = r^2\sin\theta dr d\theta d\varphi$, i.e., the scalar product $\langle \vec{\Psi} \mid \widehat{p}_r\vec{\Phi}\,\rangle$ is

$$-i\hbar\int_{r=0}^{\infty}\int_{\theta=0}^{\pi}\int_{\varphi=0}^{2\pi} \Psi^*(r,\theta,\varphi)\Big(\frac{\partial}{\partial r}+\frac{1}{r}\Big)\Phi(r,\theta,\varphi)\,dx^3.$$

Spherical harmonics are functions in $L^2(\mathcal{S}_u)$ and they are normalised by integrating with respect to the volume element $\sin\theta d\theta d\varphi$ on $\mathcal{S}_u$, as shown in Eqs. (16.43) and (16.44). It follows that

$$\langle \vec{\Psi} \mid \widehat{p}_r\vec{\Phi}\,\rangle = \int_0^{\infty} \psi(r)^*\Big\{-i\hbar\Big(\frac{\partial}{\partial r}+\frac{1}{r}\Big)\phi(r)\Big\}\,r^2\,dr.$$

---

[5] *See* Eq. (16.44). In spherical coordinates the scalar product is given by

$$\langle \vec{\Psi} \mid \widehat{p}_r\vec{\Phi}\,\rangle = -i\hbar\int_0^{\infty}\int_0^{2\pi}\int_0^{\pi} \Psi(r,\theta,\varphi)^*\Big(\frac{\partial}{\partial r}+\frac{1}{r}\Big)\Phi(r,\theta,\varphi)r^2\sin\theta dr d\theta d\varphi.$$

[6] We assume that $\phi(r)$ and $\psi(r)$ are differentiable with respect to r, i.e., they are absolutely continuous in r. *See* Wan pp. 174–175 for more details.

Using integration in parts we get

$$\langle \vec{\Psi} \mid \widehat{p}_r \vec{\Phi} \rangle = -i\hbar \int_0^\infty \psi(r)^* \frac{\partial \phi(r)}{\partial r} r^2\, dr - i\hbar \int_0^\infty \psi(r)^* \frac{1}{r} \phi(r) r^2\, dr$$

$$= -i\hbar \left[\psi^*(r)\phi(r) r^2\right]_0^\infty + i\hbar \int_0^\infty \frac{\partial(\psi(r)^* r^2)}{\partial r} \phi(r) dr$$

$$-i\hbar \int_0^\infty \psi(r)^* \frac{1}{r} \phi(r) r^2\, dr.$$

The first term vanishes due to the boundary condition $\lim_{r\to 0} r\phi(r) = 0$ and $\lim_{r\to 0} r\psi(r) = 0$ at the origin and at $r = \infty$.[7] As a result we get

$$\langle \vec{\Psi} \mid \widehat{p}_r \vec{\Phi} \rangle = i\hbar \int_0^\infty \frac{\partial(\psi(r)^* r^2)}{\partial r} \phi(r) dr - i\hbar \int_0^\infty \psi(r)^* \frac{1}{r} \phi(r) r^2\, dr$$

$$= i\hbar \int_0^\infty \left(\frac{\partial \psi(r)^*}{\partial r} r^2 + \psi(r)^* 2r\phi(r) \right) dr$$

$$-i\hbar \int_0^\infty \psi(r)^* r\phi(r)\, dr$$

$$= i\hbar \int_0^\infty \left(\frac{\partial \psi(r)^*}{\partial r} r^2 + \psi(r)^* r\phi(r) \right) dr$$

$$= \int_0^\infty \left(-i\hbar \left(\frac{\partial}{\partial r} + \frac{1}{r} \right) \psi(r) \right)^* \phi(r)\, r^2 dr$$

$$= \langle \widehat{p}_r \vec{\Psi} \mid \vec{\Phi} \rangle.$$

This shows that for $\widehat{p}_r$ to be symmetric it must be defined on a domain of vectors $\vec{\phi}$ which correspond to functions in $L^2(I\!R^3)$ which are differentiable with respect to r and satisfy the boundary condition given by Eq. (∗) in Q27(14) at the origin $r = 0$. Since the differential expression for the operator can also meaningfully act on functions $\psi(r)$ not satisfying the boundary condition at the origin we can see that the adjoint of the operator would have a bigger domain than that of $\widehat{p}_r$. So, the operator can be symmetric but not selfadjoint.

Q27(15) Explain why the product of operators $\widehat{L}(C_a)$ and $\widehat{\theta}(C_a)$ in $\vec{L}^2(C_a)$, i.e., the operator $\widehat{L}(C_a)\widehat{\theta}(C_a)$, cannot operate on the

[7] We have $\lim_{r\to\infty} r\phi(r) = 0$ and $\lim_{r\to\infty} r\psi(r) = 0$ at infinity since $\phi(r)$ and $\psi(r)$ are square-integrable with respect to the volume element $r^2 dr$ over the range $(0, \infty)$.

eigenvectors $\vec{\varphi}_n(C_a)$ of $\widehat{L}(C_a)$ in Eq. (19.36), and that[8]

$$[\widehat{\theta}(C_a), \widehat{L}(C_a)]\,\vec{\varphi}_n(C_a)$$

is not defined.[9]

SQ27(15) The situation here is similar to the discussion in §27.10.3 and in Q22(2) in Exercises and Problems for Chapter 22. Here we have

$$[\widehat{\theta}(C_a), \widehat{L}(C_a)]\,\vec{\varphi}_n(C_a) = \widehat{\theta}(C_a)\widehat{L}(C_a)\vec{\varphi}_n(C_a) - \widehat{L}(C_a)\widehat{\theta}(C_a)\vec{\varphi}_n(C_a).$$

The vector $\widehat{\theta}(C_a)\vec{\varphi}_n := \theta\varphi_n(\theta)$ does not satisfy the periodic boundary condition in Eq. (17.37).[10] This means that $\widehat{\theta}(C_a)\vec{\varphi}_n(C_a)$ is not in the domain of $\widehat{L}(C_a)$. Hence the commutator is not defined on $\vec{\varphi}_n(C_a)$.

---

[8] *See* Eqs. (17.23) and (27.111) for the definitions of $\widehat{\theta}(C_a)$ and $\widehat{L}(C_a)$.

[9] Fano pp. 407–408. *See* also Q22(2).

[10] $\varphi_n(\theta)$ is the exponential function on the right-hand side of Eq. (19.36).

Chapter 28

States, Observables and Probability Distributions

Q28(1) An electron spin is in state α_x^s.[1] Find the probability of a measurement of the z-component spin resulting in the value $\hbar/2$.

SQ28(1) The probability is given by Postulate 28.1(PDDO), i.e.,

$$\wp^{\hat{S}_z}(\alpha_x^s, \hbar/2) = \langle \vec{\alpha}_x \mid \widehat{P}_{\vec{\alpha}_z} \vec{\alpha}_x \rangle = 1/2.$$

We have used Eq. (14.25) to express $\vec{\alpha}_x$ in terms of $\vec{\alpha}_z$ and $\vec{\beta}_z$.

Q28(2) Using Eqs. (20.68) and (20.69), show that the probability distribution function and the probability measure of a proposition (as a discrete observable) represented by projector $\widehat{P}$ in state vector $\vec{\phi}$ are given by

$$\mathcal{F}^{\hat{P}}(\vec{\phi}, \tau) = \begin{cases} 0 & \tau < 0 \\ 1 - \langle \vec{\phi} \mid \widehat{P}\vec{\phi} \rangle & 0 \leq \tau < 1 \\ 1 & \tau \geq 1 \end{cases}.$$

$$\mathcal{M}^{\hat{P}}(\vec{\phi}, \Lambda) = \begin{cases} 1 - \langle \vec{\phi} \mid \widehat{P}\vec{\phi} \rangle & \text{if } \Lambda = \{0\} \\ \langle \vec{\phi} \mid \widehat{P}\vec{\phi} \rangle & \text{if } \Lambda = \{1\} \\ 0 & \text{if } \Lambda \text{ does not contain 0 or 1.} \end{cases}$$

[1] *See* §14.1.1 and §36.3 for the theory for electron spin.

Quantum Mechanics: Problems and Solutions
K. Kong Wan

ISBN 978-981-4800-72-3 (Paperback), 978-0-429-29647-5 (eBook)
www.jennystanford.com

SQ28(2) By Postulate 28.1(PDDO) or Postulate 28.2(PD) in C28.2(5) the probability distribution function $\mathcal{F}^{\hat{P}}(\vec{\phi}, \tau)$ for a proposition in state ϕ^s is given by Eq. (28.9) in terms of the spectral function $\widehat{F}^{\hat{P}}(\tau)$ of the projector $\widehat{P}$. The spectral function $\widehat{F}^{\hat{P}}(\tau)$ is given by Eq. (20.68). We get

$$\mathcal{F}^{\hat{P}}(\vec{\phi}, \tau) = \langle \vec{\phi} \mid \widehat{F}^{\hat{P}}(\tau)\vec{\phi}\,\rangle$$

$$= \begin{cases} 0 & \tau < 0 \\ 1 - \langle \vec{\phi} \mid \widehat{P}\,\vec{\phi}\,\rangle & 0 \le \tau < 1\,. \\ 1 & \tau \ge 1 \end{cases}$$

The spectral measure $\widehat{M}^{\hat{P}}(\Lambda)$ of $\widehat{P}$ is given by Eq. (20.69). The corresponding probability measure is

$$\mathcal{M}^{\hat{P}}(\vec{\phi}, \Lambda) = \langle \vec{\phi} \mid \widehat{M}^{\hat{P}}(\Lambda)\vec{\phi}\,\rangle$$

$$= \begin{cases} \langle \vec{\phi} \mid (\widehat{I\!I} - \widehat{P}\,)\vec{\phi}\,\rangle & \text{if } \Lambda = \{0\}, \\ \langle \vec{\phi} \mid \widehat{P}\vec{\phi}\,\rangle & \text{if } \Lambda = \{1\}, \\ \langle \vec{\phi} \mid \widehat{0}\vec{\phi}\rangle & \text{if } \Lambda \text{ does not contain 0 or 1.} \end{cases}$$

$$= \begin{cases} 1 - \langle \vec{\phi} \mid \widehat{P}\vec{\phi}\,\rangle & \text{if } \Lambda = \{0\}, \\ \langle \vec{\phi} \mid \widehat{P}\vec{\phi}\,\rangle & \text{if } \Lambda = \{1\}, \\ 0 & \text{if } \Lambda \text{ does not contain 0 or 1.} \end{cases}$$

Q28(3) For a particle in circular motion, the Hamiltonian $\widehat{H}(C_a)$ is given by Eq. (27.120). Show that the eigenvalues of the Hamiltonian is degenerate with eigenvectors $\vec{\varphi}_n$ given by Eq. (19.36). Write down the spectral decomposition of the Hamiltonian in the form of Eq. (20.20).

SQ28(3) The Hamiltonian $\widehat{H}(C_a)$ shares the same eigenvectors as the momentum operator $\widehat{p}(C_a)$ but with different eigenvalues, i.e.,

$$\widehat{H}(C_a)\vec{\varphi}_n(C_a) = E_n\vec{\varphi}_n(C_a), \quad E_n = \frac{1}{2m}p_n^2(C_a) = \frac{1}{2ma^2}(n\hbar)^2.$$

The ground state, i.e., $\vec{\varphi}_{n=0}(C_a)$ is nondegenerate while all other eigenvalues are degenerate with degeneracy 2 since $E_n = E_{-n}$. The spectral decomposition of $\widehat{H}(C_a)$ in terms of the projectors generated by $\vec{\varphi}_n(C_a)$ is given by Eq. (20.20), i.e.,

$$\widehat{H}(C_a) = \sum_n E_n\,\widehat{P}^{\widehat{H}(C_a)}(E_n), \quad n = 0, 1, 2, \cdots,$$

where

$$\widehat{P}^{\widehat{H}(C_a)}(E_0) = |\vec{\varphi}_0(C_a)\rangle\langle\vec{\varphi}_0(C_a)|,$$

$$\widehat{P}^{\widehat{H}(C_a)}(E_n) = |\vec{\varphi}_n(C_a)\rangle\langle\vec{\varphi}_n(C_a)| + |\vec{\varphi}_{-n}(C_a)\rangle\langle\vec{\varphi}_{-n}(C_a)|, \quad n \geq 1.$$

Q28(4) A pair of annihilation and creation operators $\widehat{a}$, $\widehat{a}^\dagger$ are defined in terms of an orthonormal basis $\{\vec{\varphi}_n, n = 0, 1, 2, \ldots\}$ in the state space $\vec{\mathcal{H}}$ of a quantum system.[2] The corresponding number operator is $\widehat{N} = \widehat{a}^\dagger\widehat{a}$.[3] The energy of the system is represented by the Hamiltonian operator $\widehat{H} = E_0\widehat{N}$.

(a) What are the possible energy values of the system?

(b) Let $\vec{\Phi}_z$ be the unit vector in Eq. (17.129). Find the probability mass function for the probability distribution of energy values of the system in state Φ^s_z described by the state vector $\vec{\Phi}_z$.

(c) Find the energy expectation values in state Φ^s_z.

SQ28(4)(a) Possible energy values are the eigenvalues of the operator $\widehat{H}$, i.e., $E_n = nE_0, n = 0, 1, 2, \cdots$, corresponding to the eigenstates defined by the eigenvectors $\vec{\varphi}_n$ of $\widehat{H}$.

SQ28(4)(b) The probability $\wp^H(\Phi^s_z, E_n)$ of a measurement obtaining the value E_n is given by

$$\wp^H(\Phi^s_z, E_n) = \langle\vec{\Phi}_z \mid \widehat{P}_{\vec{\varphi}_n}\vec{\Phi}_z\rangle,$$

where $\widehat{P}_{\vec{\varphi}_n} = |\widehat{P}_{\vec{\varphi}_n}\rangle\langle\widehat{P}_{\vec{\varphi}_n}|$ is the projector generated by the eigenvector $\vec{\varphi}_n$. Since

$$\widehat{P}_{\vec{\varphi}_n}\vec{\Phi}_z = \exp\Big(-\frac{1}{2}|z|^2\Big)\frac{z^n}{\sqrt{n!}}\vec{\varphi}_n,$$

we have

$$\begin{aligned}\wp^H(\Phi^s_z, E_n) &= \langle\vec{\Phi}_z \mid \exp\Big(-\frac{1}{2}|z|^2\Big)\frac{z^n}{\sqrt{n!}}\vec{\varphi}_n\rangle \\ &= \exp\Big(-\frac{1}{2}|z|^2\Big)\frac{z^n}{\sqrt{n!}}\langle\vec{\Phi}_z \mid \vec{\varphi}_n\rangle \\ &= \Big(\exp\Big(-\frac{1}{2}|z|^2\Big)\frac{z^n}{\sqrt{n!}}\Big)\Big(\exp\Big(-\frac{1}{2}|z|^2\Big)\frac{z^{*n}}{\sqrt{n!}}\Big) \\ &= \exp\big(-|z|^2\big)\frac{|z|^{2n}}{n!} = \exp\big(-|z|^2\big)\frac{\big(|z|^2\big)^n}{n!}.\end{aligned}$$

---

[2] *See* Definitions 17.10(1) and 17.10(2) in §17.10.

[3] *See* Definition 19.1(4).

In accordance with Theorem 22.2(1) these values define a probability mass function $\wp^{\hat{H}}(\vec{\Phi}_z, E_n)$ on the spectrum $sp_d(H) = \{E_n\}$ of $\widehat{H}$, i.e., $\wp^H(\Phi^s_z, E_n) = \wp^{\hat{H}}(\vec{\Phi}_z, E_n)$.

As a consistency check we have

$$\sum_{n=0}^{\infty} \wp^H(E_n) = \exp\left(-|z|^2\right) \sum_{n=0}^{\infty} \frac{\left(|z|^2\right)^n}{n!}$$
$$= \exp\left(-|z|^2\right) \exp\left(|z|^2\right) = 1.$$

SQ28(4)(c) The energy expectation value is

$$\sum_{n=0}^{\infty} \wp^H(E_n) E_n = \sum_{n=0}^{\infty} \exp\left(-|z|^2\right) \frac{\left(|z|^2\right)^n}{n!} n E_0$$
$$= E_0 \exp\left(-|z|^2\right) \sum_{n=1}^{\infty} \frac{\left(|z|^2\right)^n}{(n-1)!}$$
$$= E_0 \exp\left(-|z|^2\right) \sum_{n=1}^{\infty} \exp \frac{\left(|z|^2\right)^{(n-1)} |z|^2}{(n-1)!}$$
$$= |z|^2 E_0 \exp\left(-|z|^2\right) \sum_{n=1}^{\infty} \frac{\left(|z|^2\right)^{n-1}}{(n-1)!}$$
$$= |z|^2 E_0 \exp\left(-|z|^2\right) \sum_{m=0}^{\infty} \frac{\left(|z|^2\right)^m}{m!}, \quad m = n-1$$
$$= |z|^2 E_0 \exp\left(-|z|^2\right) \exp\left(|z|^2\right) = |z|^2 E_0.$$

We have used the result

$$\exp x = \sum_{n=0}^{\infty} \frac{x^m}{m!} = \sum_{n=1}^{\infty} \frac{x^{(n-1)}}{(n-1)!}.$$

Q28(5) The Fourier transform $\tilde{\varphi}(p)$ of a normalised wave function $\varphi(x)$ of a particle is

$$\tilde{\varphi}(p) = \begin{cases} 0 & p \leq -p_0, \\ 1/\sqrt{2p_0} & p \in (-p_0, p_0], \\ 0 & p \geq p_0, \end{cases}$$

where $p_0 \in I\!R$. What is the probability of a momentum measurement resulting in a value in the range $(-p_0, p_0]$? What is the momentum expectation value and uncertainty? Write down the position probability density function in terms of the inverse Fourier transform of $\tilde{\varphi}(p)$ in Eq. (18.73).

SQ28(5)[4] In accordance with Eq. (28.19) the probability of a momentum measurement the state φ^s a value in the range $(\tau_1 = -p_0, \tau_2 = p_0]$ is given by

$$\wp^p\big(\varphi^s, (\tau_1, \tau_2]\big) = \int_{\tau_1}^{\tau_2} \omega^p(\varphi^s, \tau)\, d\tau = \int_{-p_0}^{p_0} |\tilde{\varphi}(p)|^2\, dp$$

$$= \int_{-p_0}^{p_0} \left| \frac{1}{\sqrt{2p_0}} \right|^2 dp = 1.$$

The momentum expectation value is calculated in the momentum representation to be

$$\mathcal{E}(p, \varphi^s) = \int_{-\infty}^{\infty} p\, |\tilde{\varphi}(p)|^2\, d\tau = \frac{1}{2p_0} \int_{-p_0}^{p_0} p\, dp = 0.$$

This is as expected since we are as likely to get a negative value as a positive value of the momentum.

The corresponding uncertainty is given by Eq. (28.21), i.e.,

$$\Delta(p, \varphi^s) = \sqrt{\langle \vec{\varphi} \mid \hat{p}^2 \vec{\varphi} \rangle - \langle \vec{\varphi} \mid \hat{p}\, \vec{\varphi} \rangle^2} = \sqrt{\langle \vec{\varphi} \mid \hat{p}^2 \vec{\varphi} \rangle}$$

$$= \left(\frac{1}{2p_0} \int_{-p_0}^{p_0} p^2\, dp \right)^{1/2} = \frac{1}{\sqrt{3}}\, p_0.$$

The inverse Fourier transform $\varphi(x)$ of $\tilde{\varphi}(p)$ is given by Eq. (18.73), i.e., we have

$$\varphi(x) = \sqrt{\frac{\hbar}{\pi p_0}}\, \frac{\sin(p_0 x/\hbar)}{x}.$$

It follows from Eq. (28.13) that the position probability density function $w^x_\phi(x)$ is given by

$$\omega^x(\varphi^s, x) = |\varphi(x)|^2 = \frac{\hbar}{\pi p_0} \left(\frac{\sin(p_0 x/\hbar)}{x} \right)^2.$$

Q28(6) Working in the momentum representation and using Eqs. (20.54) and (20.57) find the probability distribution function for the

---

[4] *See* SQ30(2)(b).

kinetic energy of a particle of half the unit mass in one-dimensional motion along the x-axis in a given state φ^s.

SQ28(6) The state space of the particle is the Hilbert space $\vec{L}^2(I\!R)$. The state vector for the state φ^s is denoted by $\vec{\varphi}$ which is a unit vector in $\vec{L}^2(I\!R)$ corresponding to a normalised function $\varphi(x) \in L^2(I\!R)$. In the momentum representation the kinetic energy operator of a particle of half the unit mass is the multiplication operator $\widehat{\underset{\sim}{K}}(I\!R) = \widehat{\underset{\sim}{p}}(I\!R)^2$ acting on the momentum space wave functions in $\vec{L}^2(I\!R)$, e.g., $\underset{\sim}{\varphi}(p)$ which is the Fourier transforms of $\varphi(x)$. The spectral function of $\widehat{\underset{\sim}{p}}(I\!R)$ as a multiplication operator in $L^2(I\!R)$ is the characteristic function $\chi_{(-\infty,\tau]}(p)$ given by Eq. (20.29). From Eqs. (20.54), (20.56) and (20.57) we obtain the spectral function of $\widehat{\underset{\sim}{p}}(I\!R)^2$, i.e.,

$$\widehat{\underset{\sim}{F}}^{\widehat{\underset{\sim}{K}}(I\!R)}(\tau) := \begin{cases} \widehat{0} & \tau < 0 \\ \chi_{[-\sqrt{\tau},\sqrt{\tau}]}(p) & \tau \geq 0 \end{cases}.$$

The probability distribution function for state φ^s is, for $\tau > 0$,

$$\mathcal{F}^{\widehat{\underset{\sim}{K}}(I\!R)}(\tau) = \langle \vec{\underset{\sim}{\varphi}} \mid \widehat{\underset{\sim}{F}}^{\widehat{\underset{\sim}{K}}(I\!R)}(\tau)\vec{\underset{\sim}{\varphi}} \rangle = \int_{-\sqrt{\tau}}^{\sqrt{\tau}} |\underset{\sim}{\varphi}(p)|^2\, dp.$$

This is as expected. The kinetic energy does not take any negative values. The probability of the kinetic energy having a value in the range $(-\infty, \tau\,]$ for $\tau > 0$ is the same as the probability of the kinetic energy having a value in the range $[0, \tau\,]$. This is the same as the momentum having a value in the range $[-\sqrt{\tau}, \sqrt{\tau}\,]$.

Chapter 29

Time Evolution

Q29(1) Using the expression for the spectral decomposition of unitary operators in Eq. (21.6) and the result in Eq. (21.19), show that Eq. (29.5) can be obtained from the unitary evolution Eq. (29.10).

SQ29(1) For a Hamiltonian $\widehat{H}$ with a purely discrete set of eigenvalues E_ℓ corresponding a complete orthonormal set of eigenvectors $\vec{\varphi}_\ell$ and eigenprojectors $\widehat{P}_{\vec{\varphi}_\ell}$ we have, by Eq. (21.6),

$$\widehat{U}(\widehat{H}, t) = e^{-i\widehat{H}t} = \sum_\ell e^{-iE_\ell t}\,\widehat{P}_{\vec{\varphi}_\ell}.$$

$$\vec{\phi}(t) = \widehat{U}(\widehat{H}, t)\vec{\phi}(0) = \sum_\ell e^{-iE_\ell t}\left(\widehat{P}_{\vec{\varphi}_\ell}\vec{\phi}(0)\right) = \sum_\ell e^{-iE_\ell t}\, c_\ell\,\vec{\varphi}_\ell, \quad c_\ell = \langle \vec{\varphi}_\ell \mid \vec{\phi}(0)\rangle.$$

Q29(2) Verify Eq. (29.12) in the Schrödinger picture and Eq. (29.33) in the Heisenberg picture.

SQ29(2) In the Schrödinger picture the expectation value $\mathcal{E}(A, \phi^s(t))$ at time t is given by $\langle\vec{\phi}(t) \mid \widehat{A}\vec{\phi}(t)\rangle$, where $\widehat{A}$ is time-

Quantum Mechanics: Problems and Solutions
K. Kong Wan

ISBN 978-981-4800-72-3 (Paperback), 978-0-429-29647-5 (eBook)
www.jennystanford.com

independent. Differentiating this with respect to t and using the Schrödinger equation we get

$$\begin{aligned}
\frac{d\mathcal{E}\big(A, \phi^s(t)\big)}{dt} &= \langle \frac{d\vec{\phi}(t)}{dt} \mid \widehat{A}\vec{\phi}(t)\,\rangle + \langle \vec{\phi}(t) \mid \widehat{A}\,\frac{d\vec{\phi}(t)}{dt}\,\rangle \\
&= \left(\langle \frac{1}{i\hbar}\,\widehat{H}\vec{\phi}(t) \mid \widehat{A}\vec{\phi}(t)\,\rangle + \langle \vec{\phi}(t) \mid \widehat{A}\,\frac{1}{i\hbar}\,\widehat{H}\vec{\phi}(t)\,\rangle \right) \\
&= \frac{1}{i\hbar}\left(-\langle \widehat{H}\,\vec{\phi}(t) \mid \widehat{A}\vec{\phi}(t)\,\rangle + \langle \vec{\phi}(t) \mid \widehat{A}\,\widehat{H}\,\vec{\phi}(t)\,\rangle \right) \\
&= \frac{1}{i\hbar}\left(-\langle \vec{\phi}(t) \mid \widehat{H}\,\widehat{A}\vec{\phi}(t)\,\rangle + \langle \vec{\phi}(t) \mid \widehat{A}\,\widehat{H}\,\vec{\phi}(t)\,\rangle \right) \\
&= \frac{1}{i\hbar}\,\langle \vec{\phi}(t) \mid [\,\widehat{A}, \widehat{H}\,]\,\vec{\phi}(t)\,\rangle.
\end{aligned}$$

In the Heisenberg picture the expectation value at time t is given by $\langle \vec{\phi} \mid \widehat{A}(t)\vec{\phi}\,\rangle$, where $\vec{\phi}$ is time-independent. Differentiating this with respect to t and using the Heisenberg equation we get

$$\langle \vec{\phi} \mid \frac{d\widehat{A}(t)}{dt}\vec{\phi}\,\rangle = \frac{1}{i\hbar}\,\langle \vec{\phi} \mid [\,\widehat{A}(t), \widehat{H}\,]\,\vec{\phi}\,\rangle.$$

For brevity we have omitted the subscripts *Sch* for Schrödinger picture quantities and *Hei* for Heisenberg quantities. We have also assumed that the initial and the evolved vectors are in the domain of relevant operators, e.g., $\vec{\phi}(t)$ is in the domain of $\widehat{A}$.

Q29(3) The Hamiltonian of a system at time $t = 0$ is given in terms of a pair of annihilation and creation operators $\widehat{a}$ and $\widehat{a}^\dagger$ by

$$\widehat{H} = \left(\widehat{a}^\dagger\widehat{a} + \frac{1}{2}\right)\hbar\omega.$$

(a) Using the method of induction and the commutation relation of $\widehat{a}$ and $\widehat{a}^\dagger$, prove

$$\widehat{H}^n\,\widehat{a} = \widehat{a}\,(\widehat{H} - \hbar\omega)^n, \quad n = 0, 1, 2, 3, \cdots. \qquad (*)$$

(b) Assuming that the time dependence of the annihilation and creation operators in the Heisenberg picture are given by the Heisenberg equation of motion in Eq. (29.20), show that

$$\widehat{a}_{Hei}(t) = \widehat{a}_{Hei}(0)\,e^{-i\omega t}, \quad \widehat{a}^\dagger_{Hei}(t) = \widehat{a}^\dagger_{Hei}(0)\,e^{i\omega t}.$$

(c) The annihilation operator in the Schrödinger picture, denoted by $\widehat{a}_{Sch}(t)$, is related to $\widehat{a}_{Hei}(t)$ by

$$\widehat{a}_{Sch}t) = e^{\widehat{H}t/i\hbar}\,\widehat{a}_{Hei}(t)\,e^{-\widehat{H}t/i\hbar}.$$

By expanding the exponential in term of a series, i.e.,[1]

$$e^{\widehat{H}t/i\hbar} = \sum_{n=0}^{\infty} \frac{1}{n!}\left(\frac{t}{i\hbar}\right)^n \widehat{H}^n,$$

and using Eq. (∗) show that

$$e^{\widehat{H}t/i\hbar}\,\widehat{a}_{Hei}(t) = \widehat{a}_{Hei}(t)\,e^{i\omega t}\,e^{\widehat{H}t/i\hbar}.$$

Hence verify explicitly that $\widehat{a}_{Sch}(t)$ is time-independent.

SQ29(3)(a) The equation $\widehat{H}^n\,\widehat{a} = \widehat{a}(\widehat{H} - \hbar\omega)^n$ can be proved by induction:

(1) The equation is trivially true for $n = 0$ since $\widehat{H}^0$ and $(\widehat{H} - \hbar\omega)^0$ are equal to the identity operator.[2]

(2) Next we can show that it is true for $n = 1$, i.e.,

$$\begin{aligned}[\widehat{a}, \widehat{H}\,] &= [\widehat{a}, \widehat{a}^\dagger\widehat{a}\,]\,\hbar\omega = [\widehat{a}, \widehat{a}^\dagger\,]\,\widehat{a}\,\hbar\omega = \widehat{a}\,\hbar\omega \\ \Rightarrow &\quad \widehat{a}\widehat{H} - \widehat{H}\widehat{a} = \widehat{a}\,\hbar\omega \\ \Rightarrow &\quad \widehat{H}\widehat{a} = \widehat{a}\,(\widehat{H} - \hbar\omega).\end{aligned}$$

(3) Finally assume the equation to be true for an $n > 1$. Then we have

$$\begin{aligned}\widehat{H}^{n+1}\,\widehat{a} &= \widehat{H}\,\widehat{H}^n\,\widehat{a} = \widehat{H}\,\widehat{a}(\widehat{H} - \hbar\omega)^n \\ &= \widehat{a}(\widehat{H} - \hbar\omega)(\widehat{H} - \hbar\omega)^n \\ &= \widehat{a}(\widehat{H} - \hbar\omega)^{n+1}.\end{aligned}$$

It follows that the equation is true for all n.

Note that this equality is valid in the Schrödinger picture and in the Heisenberg picture for all times since the unitary transformation which relates Schrödinger picture and Heisenberg picture also preserves commutation relations.

---

[1] Assuming an appropriate domain on which the expansion is valid.

[2] *See* Definition 19.5(2) for $\widehat{A}^0 = \widehat{I\!I}$.

SQ29(3)(b) The annihilation and creation operators at time t in the Heisenberg and the Schrödinger pictures $\hat{a}_{Hei}(t)$, $\hat{a}^\dagger_{Hei}(t)$ and $\hat{a}_{Sch}(t)$, $\hat{a}^\dagger_{Sch}(t)$ are unitarily related with

$$\hat{a}_{Hei}(0) = \hat{a}_{Sch}(0) = \hat{a}_{Sch}(t) = \hat{a},$$
$$\hat{a}^\dagger_{Hei}(0) = \hat{a}^\dagger_{Sch}(0) = \hat{a}^\dagger_{Sch}(t) = \hat{a}^\dagger.$$

The operators in the Heisenberg picture are obtainable from the corresponding quantities in the Schrödinger picture by a unitary transformation given by Eq. (29.13), i.e.,

$$\hat{a}_{Hei}(t) = \hat{U}(t)^\dagger \hat{a}_{Hei}(0)\hat{U}(t),$$
$$\hat{a}^\dagger_{Hei}(t) = \hat{U}(t)^\dagger \hat{a}^\dagger_{Hei}(0)\hat{U}(t),$$

where $\hat{U}(t) = \exp(-i\hat{H}t)$. Since unitary transformations preserve commutation relations we have

$$[\hat{a}_{Hei}(t), \hat{a}^\dagger_{Hei}(t)] = [\hat{a}_{Sch}(t), \hat{a}^\dagger_{Sch}(t)] = [\hat{a}_{Sch}(0), \hat{a}^\dagger_{Sch}(0)] = 1.$$

In accordance with Eq. (29.15) the Hamiltonian in the Heisenberg picture at time t is expressible in terms of $\hat{a}_{Hei}(t)$, and $\hat{a}^\dagger_{Hei}(t)$ as[3]

$$\hat{H}(t) = \hat{U}(t)^\dagger \hat{H}(0)\hat{U}(t) = \hbar\omega\left(\hat{a}^\dagger_{Hei}(t)\hat{a}_{Hei}(t) + \frac{1}{2}\right).$$

It follows that

$$\begin{aligned}[\hat{a}_{Hei}(t), \hat{H}(t)] &= \left[\hat{a}_{Hei}(t), \hbar\omega\left(\hat{a}^\dagger_{Hei}(t)\hat{a}_{Hei}(t) + \frac{1}{2}\right)\right] \\ &= \hbar\omega\,[\hat{a}_{Hei}(t), \hat{a}^\dagger_{Hei}(t)\hat{a}_{Hei}(t)] \\ &= \hbar\omega\,\hat{a}_{Hei}(t).\end{aligned}$$

The Heisenberg equation for $\hat{a}_{Hei}(t)$ is

$$i\hbar\frac{d}{dt}\hat{a}_{Hei}(t) = [\hat{a}_{Hei}(t), \hat{H}(t)] = \hbar\omega\,\hat{a}_{Hei}(t)$$
$$\Rightarrow \quad \frac{d}{dt}\hat{a}_{Hei}(t) = -i\omega\,\hat{a}_{Hei}(t).$$

The solution is clearly

$$\hat{a}_{Hei}(t) = \hat{a}_{Hei}(0)e^{-i\omega t},$$

---

[3]The Hamiltonian is time-independent, even though it is written as $\hat{H}(t)$, i.e.,we have $\hat{H}(t) = \hat{H}(0) = \hat{H}$, as explicitly shown in Q29(3)(c).

where $\widehat{a}_{Hei}(0)$ can be identified with the corresponding Schrödinger picture operator $\widehat{a}_{Sch}(0)$. The time dependence of $\widehat{a}^{\dagger}_{Hei}(t)$ is

$$\widehat{a}^{\dagger}_{Hei}(t) = \left(\widehat{a}_{Hei}(0)e^{-i\omega t}\right)^{\dagger} = \widehat{a}^{\dagger}_{Hei}(0)e^{i\omega t}.$$

SQ29(3)(c) The annihilation operator in the Schrödinger picture, denoted by $\widehat{a}_{Sch}(t)$, is related to $\widehat{a}_{Hei}(t)$ by

$$\widehat{a}_{Sch}(t) = e^{\widehat{H}t/i\hbar}\,\widehat{a}_{Hei}(t)\,e^{-\widehat{H}t/i\hbar}.$$

Our problem is to work out the right-hand side and show explictly that $\widehat{a}_{Sch}(t)$ is time-independent. To work out the right-side we first consider[4]

$$\begin{aligned}
&e^{\widehat{H}t/i\hbar}\,\widehat{a}_{Hei}(t)\\
&= \sum_{n=0}^{\infty}\frac{1}{n!}\left(\frac{\widehat{H}\,t}{i\hbar}\right)^{n}\widehat{a}_{Hei}(t)\\
&= e^{-i\omega t}\sum_{n=0}^{\infty}\frac{1}{n!}\left(\frac{\widehat{H}\,t}{i\hbar}\right)^{n}\widehat{a}_{Hei}(0)\\
&= e^{-i\omega t}\sum_{n=0}^{\infty}\frac{1}{n!}\left(\frac{t}{i\hbar}\right)^{n}\widehat{H}^{n}\,\widehat{a} \quad \text{since } \widehat{a}_{Hei}(0) = \widehat{a}\\
&= e^{-i\omega t}\sum_{n=0}^{\infty}\frac{1}{n!}\left(\frac{t}{i\hbar}\right)^{n}\widehat{a}(\widehat{H} - \hbar\omega)^{n} \quad \text{by Eq. } (*) \text{ in Q29(3)(a)}\\
&= e^{-i\omega t}\,\widehat{a}\sum_{n=0}^{\infty}\frac{1}{n!}\left(\frac{(\widehat{H} - \hbar\omega)\,t}{i\hbar}\right)^{n}\\
&= e^{-i\omega t}\,\widehat{a}\,e^{(\widehat{H}-\hbar\omega)t/i\hbar} = \widehat{a}\,e^{\widehat{H}t/i\hbar}.
\end{aligned}$$

It follows that

$$e^{\widehat{H}t/i\hbar}\,\widehat{a}_{Hei}(t)\,e^{-\widehat{H}t/i\hbar} = \widehat{a} = \widehat{a}_{Sch}(t),$$

which is time-independent as expected.

Q29(4) Let $\widehat{A}$ be the selfadjoint operator representing an observable A in the Schrödinger picture and let $\widehat{F}^{\hat{A}}(\tau)$ be its spectral function. Let the corresponding operator at time t in the Heisenberg picture be denoted by $\widehat{A}_{Hei}(t)$ and let $\widehat{F}^{\hat{A}_{Hei}(t)}(\tau)$ be its spectral

[4] Making used of Eq. (29.19) which relate $\widehat{a}_{Hei}(t)$ to $\widehat{a}_{Hei}(0)$.

function. How are these two spectral functions related and how are the probability distribution functions generated by these spectral functions in a given state related?

SQ29(4) At time $t = 0$ let the state vector be $\vec{\phi}$ and the operator for the observable in question be $\widehat{A}$. At time $t > 0$ let the evolved state vector be denoted by $\vec{\phi}(t)$ in the Schrödinger picture. In the Heisenberg picture the evolved vector $\vec{\phi}_{Hei}(t)$ at time t remains $\vec{\phi}$, i.e., we have $\vec{\phi}_{Hei}(t) = \vec{\phi}$. So, we can simply rewrite $\vec{\phi}_{Hei}(t)$ as $\vec{\phi}$ without the variable t. The two vectors $\vec{\phi}(t)$ and $\vec{\phi}_{Hei}$ are related by Eq. (29.36), i.e., we have by $\vec{\phi}(t) = \widehat{U}(t)\vec{\phi}_{Hei}$. The evolved operator remains $\widehat{A}$ in the Schrödinger picture, while in the Heisenberg picture the evolved operator becomes $\widehat{A}_{Hei}(t) = \widehat{U}(t)^{\dagger}\widehat{A}\widehat{U}(t)$ as shown in Eq. (29.39). For brevity of notation we have omitted the subscripts Sch for quantities in the the Schrödinger picture.

In accordance of Theorem 15.3(2) the spectral functions $\widehat{F}^{\hat{A}}(\tau)$ and $\widehat{F}^{\hat{A}_{Hei}(t)}(\tau)$ of $\widehat{A}$ and $\widehat{A}_{Hei}(t)$ are related by the unitary transformation generated by the unitary operator $\widehat{U}(t)$ in the same way the operators of the two pictures are related by Eq. (29.43), i.e.,

$$\widehat{F}^{\hat{A}}(\tau) = \widehat{U}(t)\widehat{F}^{\hat{A}_{Hei}(t)}(\tau)\widehat{U}(t)^{\dagger}.$$

The probability distribution function in the Schrödinger picture in a given state at time t is given by

$$\mathcal{F}^{\hat{A}}(\vec{\phi}(t), \tau) = \langle\vec{\phi}(t) \mid \widehat{F}^{\hat{A}}(\tau)\vec{\phi}(t)\rangle.$$

The probability distribution function in the Heisenberg picture in the same given state at time t is given by

$$\mathcal{F}^{\hat{A}_{Hei}(t)}(\vec{\phi}_{Hei}, \tau) = \langle\vec{\phi}_{Hei} \mid \widehat{F}^{\hat{A}_{Hei}(t)}(\tau)\vec{\phi}_{Hei}\rangle.$$

Then we have

$$\begin{aligned}\mathcal{F}^{\hat{A}}(\vec{\phi}(t), \tau) &= \langle\widehat{U}(t)\vec{\phi}_{Hei} \mid \left(\widehat{U}(t)\widehat{F}^{\hat{A}_{Hei}(t)}(\tau)\widehat{U}(t)^{\dagger}\right)\widehat{U}(t)\vec{\phi}_{Hei}\rangle \\ &= \langle\widehat{U}(t)\vec{\phi}_{Hei} \mid \widehat{U}(t)\widehat{F}^{\hat{A}_{Hei}(t)}(\tau)\vec{\phi}_{Hei}\rangle \\ &= \langle\vec{\phi}_{Hei} \mid \widehat{F}^{\hat{A}_{Hei}(t)}(\tau)\vec{\phi}_{Hei}\rangle = \mathcal{F}^{\hat{A}_{Hei}(t)}(\vec{\phi}, \tau).\end{aligned}$$

This is as expected since the probability distribution function is measurable and it should not depend on any particular mathematical description.

Q29(5) The Hamiltonian of a quantum particle of mass m in one-dimensional motion along the x-axis with potential energy $V(\hat{x})$ is given by

$$\hat{H} = \frac{1}{2m}\hat{p}^2 + V(\hat{x}).$$

Assuming $V(\hat{x})$ to be a polynomial function of $\hat{x}$ show that in the Heisenberg picture, we have[5]

$$m\frac{d}{dt}\langle \vec{\phi} \mid \hat{x}\vec{\phi}\,\rangle = \langle \vec{\phi} \mid \hat{p}\vec{\phi}\,\rangle,$$

$$\frac{d}{dt}\langle \vec{\phi} \mid \hat{p}\vec{\phi}\,\rangle = -\langle \vec{\phi} \mid \frac{dV(\hat{x})}{d\hat{x}}\vec{\phi}\,\rangle.$$

Establish the following **Ehrenfest's theorem**:

$$m\frac{d^2}{dt^2}\langle \vec{\phi} \mid \hat{x}\vec{\phi}\,\rangle = \langle \vec{\phi} \mid \hat{F}\vec{\phi}\,\rangle, \quad \text{where } \hat{F} = -\frac{dV(\hat{x})}{d\hat{x}}.$$

Discuss the physical significance of this result,

SQ29(5) From Eq. (29.32) we get[6]

$$\frac{d}{dt}\hat{x}(t) = \frac{\partial \hat{H}(t)}{\partial \hat{p}(t)} = \frac{1}{m}\hat{p}(t),$$

$$\frac{d}{dt}\hat{p}(t) = -\frac{\partial \hat{H}(t)}{\partial \hat{x}(t)} = -\frac{\partial V(\hat{x}(t))}{\partial \hat{x}}.$$

It follows that

$$\frac{d^2}{dt^2}\hat{x}(t) = \frac{1}{m}\frac{d}{dt}\hat{p}(t) \Rightarrow m\frac{d^2}{dt^2}\hat{x}(t) = -\frac{\partial V(\hat{x}(t))}{\partial \hat{x}}.$$

Taking the expectation values on both sides we get

$$m\frac{d^2}{dt^2}\langle \vec{\phi} \mid \hat{x}(t)\vec{\phi}\,\rangle = \frac{d}{dt}\langle \vec{\phi} \mid \hat{F}\vec{\phi}\,\rangle,$$

where

$$\hat{F} = -\frac{dV(\hat{x}(t))}{d\hat{x}(t)},$$

which is the Ehrenfest's theorem. This is the equivalence of Newton's second law in classical mechanics. The significance here is that it

[5]The subscript *Hei* for Heisenberg picture quantities is omitted for brevity. We also assume that $\vec{\phi}$ is in an appropriate domain of all the operators involved.

[6]The subscript *Hei* for Heisenberg picture quantities is omitted for brevity.

is the second order time derivative of the expectation value of the position which is related to the expectation value of the force.

Q29(6) Discuss the fundamental differences in the concept of constants of motion in classical and quantum mechanics.

SQ29(6) In classical mechanics a constant of motion is an observable whose value mains constant and independent of time. A quantum observable is a constant of motion if its expectation value in a given state is a constant and independent of time. Bearing in mind that in a given state the observable may not possess a value it is not meaningful to say that the observable would have a definite initial value and as time evolves the observable would still possess this value. In the Schrödinger picture let the initial state be $\vec{\phi}(0)$ and the final state be $\vec{\phi}(T)$. Then an initial measured value in state described $\vec{\phi}(0)$ may well not be the same as an individual measured value of the observable in state described $\vec{\phi}(T)$ at a later time T. But the expectation value of the observable at time T is the same as that at time 0.

Chapter 30

On States after Measurement

Q30(1) An electron spin is in state α^s_x.[1] What is the state and the state vector immediately after a measurement of the z-component spin resulting in the value $\hbar/2$?

SQ30(1) Immediately after a measurement of the z-component spin resulting in the value $\hbar/2$ state is α^s_z by Postulate 30.1.1(PPDO). The corresponding state vector is $\vec{\alpha}_z$.

Q30(2) Find the subspace associated with the projector $\widehat{F}^{\hat{x}}(x_2) - \widehat{F}^{\hat{x}}(x_1)$ on $\vec{L}^2(I\!R)$.

SQ30(2) The subspace consists of all the vectors $\vec{\varphi}$ in $\vec{L}^2(I\!R)$ satisfying the equation

$$\left(\widehat{F}^{\hat{x}}(x_2) - \widehat{F}^{\hat{x}}(x_1)\right)\vec{\varphi} = \vec{\varphi}.$$

In terms of functions in $L^2(I\!R)$ this equation becomes

$$\chi_{(x_1,x_2]}(x)\varphi(x) = \varphi(x).$$

The solutions to this equation are all the functions in $L^2(I\!R)$ which vanish outside the interval $\Lambda = (x_1, x_2]$. This subspace is identifiable with $\vec{L}^2(\Lambda)$.

[1] *See* §14.1.1 for the theory for electron spin.

Quantum Mechanics: Problems and Solutions
K. Kong Wan

ISBN 978-981-4800-72-3 (Paperback), 978-0-429-29647-5 (eBook)
www.jennystanford.com

Q30(3) Consider a particle in one-dimensional motion along the x-axis. According to Postulate 30.2.1(PPCO), if a momentum measurement yields a value in the interval $(p_1, p_2]$ the state vector $\vec{\varphi}$ in the momentum representation right after the measurement must satisfy

$$\left(\widehat{F}^{\hat{p}}_{\sim}(p_2) - \widehat{F}^{\hat{p}}_{\sim}(p_1)\right)\vec{\varphi} = \vec{\varphi}, \qquad (*)$$

where $\widehat{F}^{\hat{p}}_{\sim}(p)$ is the spectral function of the momentum operator given in Theorem 20.4.2(1), i.e., $\widehat{F}^{\hat{p}}_{\sim}(p)$ is defined by the characteristic function $\underset{\sim}{\chi}_{(-\infty,\tau]}(p)$ of the interval $(-\infty, \tau]$ on the momentum space $\underset{\sim}{I\!R}$.

Express Eq. $(*)$ in terms of $\underset{\sim}{\chi}_{(-\infty,\tau]}(p)$ and functions $\underset{\sim}{\varphi}(p)$ in $L^2(\underset{\sim}{I\!R})$. Give an example of such a function in the momentum space and its corresponding function in the coordinate space.

SQ30(3) In terms of $\underset{\sim}{\chi}_{(-\infty,\tau]}(p)$ and $\underset{\sim}{\varphi}(p)$ Eq. $(*)$ becomes

$$\left(\underset{\sim}{\chi}_{(-\infty,\, p_2]}(p) - \underset{\sim}{\chi}_{(-\infty,\, p_1]}(p)\right)\underset{\sim}{\varphi}(p) = \underset{\sim}{\varphi}(p).$$

Since

$$\underset{\sim}{\chi}_{(-\infty,\, p_2]}(p) - \underset{\sim}{\chi}_{(-\infty,\, p_1]}(p) = \underset{\sim}{\chi}_{(p_1,\, p_2]}(p),$$

the equation becomes

$$\underset{\sim}{\chi}_{(p_1,\, p_2]}(p)\underset{\sim}{\varphi}(p) = \underset{\sim}{\varphi}(p).$$

This is similar to the situation in position measurement discussed in SQ30(2). Clearly the following normalised function

$$\underset{\sim}{\varphi}(p) = \begin{cases} 1/\sqrt{p_2 - p_1} & p \in (p_1,\, p_2] \\ 0 & p \notin (p_1,\, p_2] \end{cases},$$

in the momentum space satisfies the above equation. The corresponding function $\varphi(x)$ in the coordinate representation space is the inverse Fourier transform of $\underset{\sim}{\varphi}(p)$, i.e.,

$$\begin{aligned} \varphi(x) &= \frac{1}{\sqrt{2\pi\hbar(p_2 - p_1)}} \int_{p_1}^{p_2} e^{ipx}\, dp \\ &= \frac{1}{\sqrt{2\pi\hbar(p_2 - p_1)}}\, \frac{e^{ip_2x/\hbar} - e^{ip_1x/\hbar}}{ix/\hbar} \\ &= \sqrt{\frac{\hbar}{2\pi(p_2 - p_1)}}\, \frac{e^{ip_2x/\hbar} - e^{ip_1x/\hbar}}{ix}. \end{aligned}$$

This function reduces to that in Eq. (18.73) when $p_2 = p_0$ and $p_1 = -p_0$.

Chapter 31

Pure and Mixed States

Q31(1) Explain why the transition shown in Eq. (31.1) due to measurement cannot be generated by a unitary transformation.[1]

SQ31(1) In the Schrödinger picture on quantum time evolution as stated in Postulate 29.1.2 (TESP) we have:

(1) If an initial state is described by a state vector the evolved state would be described by the evolved state vector in Eq. (29.10), since a unitary transformation will yield a unique transformed vector at any given time t from a given vector. A pure state would evolves into another pure state.

(2) If the initial state is represented by a one-dimensional projector then the evolved state is again represented by a one-dimensional projector since the unitary transform of a one-dimensional projector is again a one-dimensional projector.[2] A pure state would evolves into another pure state.

[1]This means that the transition as a time evolution does not satisfy Postulate 29.1.2(TESP) which applies to quantum evolution in the Schrödinger picture.
[2]*See* Q18(9).

Quantum Mechanics: Problems and Solutions
K. Kong Wan

ISBN 978-981-4800-72-3 (Paperback), 978-0-429-29647-5 (eBook)
www.jennystanford.com

The transition in Eq. (31.1) shows a transition from a pure state to a classical mixture of pure states. This contradicts what has been said above. Hence such a transition cannot be a unitary transformation of the initial pure state. We can also discuss the situation in terms of state vectors or one-dimensional projectors:

(1) The transition shown in Eq. (31.1) means that an initial state vector is transformed into many different final state vectors. It follows that the transition cannot be a unitary transformation from the initial state vector.

(2) A one-dimensional projector cannot evolve unitarily to a density operator which is not a projector.

Q31(2) Prove Eqs. (31.6) and (31.7).

SQ31(2) To prove Eq. (31.6) we can choose an orthonormal basis $\{\vec{\varphi}_\ell\}$ such that $\vec{\varphi}_1 = \vec{\phi}$. Then $\vec{\phi}$ will be orthogonal to all $\vec{\varphi}_\ell$ for $\ell \geq 2$ so that $\widehat{P}_{\vec{\phi}}\, \vec{\varphi}_\ell = \vec{0}$ except for $\ell = 1$. Then

$$\begin{aligned}\operatorname{tr}\big(\widehat{B}\widehat{P}_{\vec{\phi}}\big) &= \sum_\ell \langle \vec{\varphi}_\ell \mid \widehat{B}\widehat{P}_{\vec{\phi}}\, \vec{\varphi}_\ell\rangle \\ &= \langle \vec{\varphi}_1 \mid \widehat{B}\widehat{P}_{\vec{\phi}}\, \vec{\varphi}_1\rangle = \langle \vec{\phi} \mid \widehat{B}\widehat{P}_{\vec{\phi}}\, \vec{\phi}\,\rangle = \langle \vec{\phi} \mid \widehat{B}\, \vec{\phi}\,\rangle.\end{aligned}$$

Equation (31.7) can be proved in the same way.

Q31(3) Prove Eqs. (31.17) and (31.18).

SQ31(3) A general density operator $\widehat{D}$ is decomposable as a convex combination of projectors in accordance with Eq. (18.22) or Eq. (18.24), i.e., $\widehat{D} = \sum_\ell \omega_\ell\, \widehat{P}_{\vec{\varphi}_\ell}$. Since the trace operation is linear the right-hand side of Eq. (31.15) becomes

$$\begin{aligned}\operatorname{tr}\big(\widehat{F}^{\widehat{B}}(\tau)\widehat{D}\big) &= \operatorname{tr}\Big(\widehat{F}^{\widehat{B}}(\tau)\sum_\ell \omega_\ell\, \widehat{P}_{\vec{\varphi}_\ell}\Big) \\ &= \sum_\ell \omega_\ell \operatorname{tr}\big(\widehat{F}^{\widehat{B}}(\tau)\widehat{P}_{\vec{\varphi}_\ell}\big).\end{aligned}$$

By the result obtained in SQ31(2) we have

$$\begin{aligned}&\operatorname{tr}\big(\widehat{F}^{\widehat{B}}(\tau)\widehat{P}_{\vec{\varphi}_\ell}\big) = \langle \vec{\varphi}_\ell \mid \widehat{F}^{\widehat{B}}(\tau)\vec{\varphi}_\ell\rangle \\ \Rightarrow\ &\operatorname{tr}\big(\widehat{F}^{\widehat{B}}(\tau)\widehat{D}\big) = \sum_\ell \omega_\ell\, \langle \vec{\varphi}_\ell \mid \widehat{F}^{\widehat{B}}(\tau)\vec{\varphi}_\ell\rangle.\end{aligned}$$

By Postulate 28.2(PD), i.e., Eq. (28.9), we have

$$\mathcal{F}^{\hat{B}}(\vec{\varphi}_\ell, \tau) = \langle \vec{\varphi}_\ell \mid \widehat{F}^{\hat{B}}(\tau)\vec{\varphi}_\ell \rangle$$
$$\Rightarrow \quad \text{tr}\,(\widehat{F}^{\hat{B}}(\tau)\widehat{D}) = \sum_\ell \omega_\ell\, \mathcal{F}^{\hat{B}}(\vec{\varphi}_\ell, \tau),$$

which is the desired Eq. (31.17).

Equation (31.18) can be proved in the same way.

Q31(4) Prove Eq. (31.20) for the expectation value.

SQ31(4) On account of Eqs. (31.15) and (31.17) the integral in Eq. (31.19) can be evaluated as follows:

$$\int_{-\infty}^{\infty} \tau\, d_\tau \big(\text{tr}\,(\widehat{F}^{\hat{B}}(\tau)\widehat{D})\big) = \int_{-\infty}^{\infty} \tau\, d_\tau \Big(\sum_\ell \omega_\ell\, \mathcal{F}^{\hat{B}}(\vec{\varphi}_\ell, \tau)\Big)$$
$$= \sum_\ell \omega_\ell \Big(\int_{-\infty}^{\infty} \tau\, d_\tau \mathcal{F}^{\hat{B}}(\vec{\varphi}_\ell, \tau)\Big) = \sum_\ell \omega_\ell \langle \vec{\varphi}_\ell \mid \widehat{B}\vec{\varphi}_\ell \rangle.$$

We have used Eq. (22.3), i.e.,

$$\int_{-\infty}^{\infty} \tau\, d_\tau \mathcal{F}^{\hat{B}}(\vec{\varphi}_\ell, \tau) = \langle \vec{\varphi}_\ell \mid \widehat{B}\vec{\varphi}_\ell \rangle.$$

On the other hand we have

$$\text{tr}\,(\widehat{B}\widehat{D}) = \text{tr}\,\Big(\widehat{B} \sum_\ell \omega_\ell\, \widehat{P}_{\vec{\varphi}_\ell}\Big)$$
$$= \sum_\ell \omega_\ell\, \text{tr}\,(\widehat{B}\widehat{P}_{\vec{\varphi}_\ell}) = \sum_\ell \omega_\ell\, \langle \vec{\varphi}_\ell \mid \widehat{B}\vec{\varphi}_\ell \rangle.$$

Equation (31.20) then follows, i.e.,

$$\mathcal{E}(\widehat{B}, \widehat{D}) = \text{tr}\,(\widehat{B}\widehat{D}) = \sum_\ell \omega_\ell\, \langle \vec{\varphi}_\ell \mid \widehat{B}\vec{\varphi}_\ell \rangle.$$

Chapter 32

Superselection Rules

Q32(1) Prove Eq. (32.4).

SQ32(1) Since $\vec{\mathcal{S}}^{(n)}$ is a complete orthogonal family of subspaces their corresponding projectors $\widehat{P}^{(n)}$ would constitute a complete orthogonal family of projectors. It follows that their sum is equal to the identity operator, i.e., $\sum_n \widehat{P}^{(n)} = \widehat{I\!I}$. Then

$$\widehat{B}_{re} = \Big(\sum_n \widehat{P}^{(n)}\Big)\widehat{B}_{re}\Big(\sum_m \widehat{P}^{(m)}\Big)$$
$$= \sum_n \widehat{P}^{(n)}\widehat{B}_{re}\widehat{P}^{(n)} + \sum_{n\neq m} \widehat{P}^{(n)}\widehat{B}_{re}\widehat{P}^{(m)}.$$

Being reducible by $\vec{\mathcal{S}}^{(n)}$ the operator $\widehat{B}_{re}$ would commute with $\widehat{P}^{(n)}$ by Theorem 17.9(1) so that the terms in the second sum above can be written as $\widehat{B}_{re}\widehat{P}^{(n)}\widehat{P}^{(m)}$. Since the projectors are orthogonal these terms vanish. We get

$$\widehat{B}_{re} = \sum_n \widehat{P}^{(n)}\widehat{B}_{re}\widehat{P}^{(n)} = \sum_n \widehat{B}^{(n)},$$

where $\widehat{B}^{(n)} = \widehat{P}^{(n)}\widehat{B}_{re}\widehat{P}^{(n)}$.

Q32(2) For the example in §32.3 discuss how the initial state $\eta^{(0)\oplus}$ given by Eq. (32.15) would evolve in the Schrödinger picture under a selfadjoint and decomposable Hamiltonian $\widehat{H}^{\oplus} = \widehat{B}^{(-)} \oplus \widehat{B}^{(0)} \oplus \widehat{B}^{(+)}$.

Quantum Mechanics: Problems and Solutions
K. Kong Wan

ISBN 978-981-4800-72-3 (Paperback), 978-0-429-29647-5 (eBook)
www.jennystanford.com

SQ32(2) The given initial state is $\vec{\eta}^{(0)\oplus} = \vec{0}^{(-)} \oplus \vec{\eta}^{(0)} \oplus \vec{0}^{(+)}$. Since the subspace $\vec{\mathcal{S}}^{(0)}$ is one-dimensional the vector $\vec{\eta}^{(0)}$ must be an eigenvector $\widehat{B}^{(0)}$ in Eq. (32.17), i.e., we have

$$\widehat{B}^{(0)}\vec{\eta}^{(0)} = b_0\vec{\eta}^{(0)} \quad \text{for some } b_0 \in I\!R.$$

For the given Hamiltonian $\widehat{H}^{\oplus} = \widehat{B}^{(-)} \oplus \widehat{B}^{(0)} \oplus \widehat{B}^{(+)}$ we have

$$\widehat{H}^{\oplus}\vec{\eta}^{(0)\oplus} = b_0\vec{\eta}^{(0)\oplus}.$$

It follows from Eq. (29.4) that in the Schrödinger picture the evolved state vector for the initial state vector $\vec{\eta}_0^{\oplus}$ is given by

$$\vec{\eta}^{(0)\oplus}(t) = e^{-i b_0 t}\vec{\eta}^{(0)\oplus}.$$

Here both the initial and the final states are pure. The state vector $\vec{\eta}^{(0)\oplus}(t)$ remains in the subspace $\vec{\mathcal{S}}^{(0)}$ for all times.

Q32(3) A system has a two-dimensional state space $\vec{\mathcal{H}}$. A superselection rule operates with the state space $\vec{\mathcal{H}}$ having the following preferred direct sum decomposition:

$$\vec{\mathcal{H}} = \vec{\mathcal{H}}^{\oplus} = \vec{\mathcal{H}}^{(1)} \oplus \vec{\mathcal{H}}^{(2)},$$

where $\vec{\mathcal{H}}^{(1)}$ is spanned by a unit vector $\vec{\eta}^{(1)} \in \vec{\mathcal{H}}$ and $\vec{\mathcal{H}}^{(2)}$ is spanned by a unit vector $\vec{\eta}^{(2)} \in \vec{\mathcal{H}}$ which is orthogonal to $\vec{\eta}^{(1)}$. Let $\widehat{L}$ be an operator on $\vec{\mathcal{H}}$ defined by

$$\widehat{L}\vec{\eta}^{(1)} = \vec{\eta}^{(2)}, \quad \widehat{L}\vec{\eta}^{(2)} = \vec{\eta}^{(1)}.$$

(a) Show that $\widehat{L}$ is selfadjoint and explain why $\widehat{L}$ cannot represent an observable.

(b) Suppose the Hamiltonian of the system is $\lambda\widehat{L}$, where $\lambda \in I\!R$. Show that in the Schrödinger picture the initial state vector $\vec{\eta}(0) = i\vec{\eta}^{(2)}$ at time $t = 0$ evolves in time to the following state vector at time t:

$$\vec{\eta}(t) = \sin(\lambda t/\hbar)\vec{\eta}^{(1)} + i\cos(\lambda t/\hbar)\vec{\eta}^{(2)}.$$

(c) Discuss how the evolution may cause a transition from a pure state to a mixture.

SQ32(3)(a) The operator $\widehat{L}$ satisfies the selfadjointness condition $\langle \vec{\eta}^{(m)} \mid \widehat{L}\vec{\eta}^{(n)} \rangle = \langle \widehat{L}\vec{\eta}^{(m)} \mid \vec{\eta}^{(n)} \rangle$ $\forall n, m = 1, 2$, since

$$\langle \vec{\eta}^{(1)} \mid \widehat{L}\vec{\eta}^{(1)} \rangle = \langle \widehat{L}\vec{\eta}^{(1)} \mid \vec{\eta}^{(1)} \rangle = 0,$$
$$\langle \vec{\eta}^{(1)} \mid \widehat{L}\vec{\eta}^{(2)} \rangle = \langle \widehat{L}\vec{\eta}^{(1)} \mid \vec{\eta}^{(2)} \rangle = 1,$$
$$\langle \vec{\eta}^{(2)} \mid \widehat{L}\vec{\eta}^{(1)} \rangle = \langle \widehat{L}\vec{\eta}^{(2)} \mid \vec{\eta}^{(1)} \rangle = 1,$$
$$\langle \vec{\eta}^{(2)} \mid \widehat{L}\vec{\eta}^{(2)} \rangle = \langle \widehat{L}\vec{\eta}^{(2)} \mid \vec{\eta}^{(2)} \rangle = 0.$$

It follows that $\widehat{L}$ is selfadjoint. By Definition 32.1(5) the operator $\widehat{L}$ cannot represent an observable since it is not decomposable.

SQ32(3)(b) First the state vector $\vec{\eta}(t)$ satisfies the initial condition, i.e., $\vec{\eta}(t = 0) = \vec{\eta}(0)$. It also satisfies the Schrödinger equation, i.e.,

$$i\hbar \frac{d\vec{\eta}(t)}{dt} = i\lambda \cos(\lambda t/\hbar)\, \vec{\eta}^{(1)} + \lambda \sin(\lambda t/\hbar)\, \vec{\eta}^{(2)}.$$
$$\widehat{H}\,\vec{\eta}(t) = \lambda \sin(\lambda t/\hbar)\, \vec{\eta}^{(2)} + i\lambda \cos b(\lambda t/\hbar)\, \vec{\eta}^{(1)}.$$
$$\Rightarrow \quad i\hbar \frac{d\vec{\eta}(t)}{dt} = \widehat{H}\,\vec{\eta}(t).$$

SQ32(3)(c) The initial state vector $\vec{\eta}(0) = i\vec{\eta}^{(2)}$ describes a pure state while the evolved state vector $\vec{\eta}(t)$ at time t represents a classical mixture of $\vec{\eta}^{(1)}$ and $\vec{\eta}^{(2)}$, except for certain specific times t_n when either $\sin(\lambda t_n/\hbar)$ or $\cos(\lambda t_n/\hbar)$ is zero. Here we have an example of an evolution from a pure state to a mixed state. This is because that the Hamiltonian is not decomposable, and it does not represent an observable of the system. The physical significance of this is discussed in §34.7.2 on quantum measurement.

Chapter 33

Many-Particle Systems

Q33(1) Calculate the expectation value of the two-particle observable represented by the operator in Eq. (33.9) in the state described by the vector $\vec{\Phi}^{(c)}$ in Eq. (33.2).

SQ33(1) We can make use of Eqs. (33.5) and (33.8). For the two-particle observable in Eq. (33.9) the expression for the expectation value in becomes

$$\langle \vec{\Phi}^{(c)} \mid \left(\widehat{A}^{(1} \otimes \widehat{II}^{(2)} + \widehat{II}^{(1)} \otimes \widehat{A}^{(2)} \right) \vec{\Phi}^{(c)} \rangle^{\otimes}$$

$$= \langle \vec{\Phi}^{(c)} \mid \left(\widehat{A}^{(1} \otimes \widehat{II}^{(2)} \right) \vec{\Phi}^{(c)} \rangle^{\otimes} + \langle \vec{\Phi}^{(c)} \mid \left(\widehat{II}^{(1)} \otimes \widehat{A}^{(2)} \right) \vec{\Phi}^{(c)} \rangle^{\otimes}$$

$$= \sum_{j,k,m,n} c_{j,k}^{*} c_{m,n} \langle \vec{\phi}_j^{(1)} \mid \widehat{A}^{(1)} \vec{\phi}_m^{(1)} \rangle^{(1)} \langle \vec{\phi}_k^{(2)} \mid \widehat{II}^{(2)} \vec{\phi}_n^{(2)} \rangle^{(2)}$$

$$+ \sum_{j,k,m,n} c_{j,k}^{*} c_{m,n} \langle \vec{\phi}_j^{(1)} \mid \widehat{II}^{(1)} \vec{\phi}_m^{(1)} \rangle^{(1)} \langle \vec{\phi}_k^{(2)} \mid \widehat{A}^{(2)} \vec{\phi}_n^{(2)} \rangle^{(2)}$$

$$= \sum_{j,k,m} c_{j,k}^{*} c_{m,k} \langle \vec{\phi}_j^{(1)} \mid \widehat{A}^{(1)} \vec{\phi}_m^{(1)} \rangle^{(1)}$$

$$+ \sum_{j,k,n} c_{j,k}^{*} c_{j,n} \langle \vec{\phi}_k^{(2)} \mid \widehat{A}^{(2)} \vec{\phi}_n^{(2)} \rangle^{(2)}.$$

Q33(2) Show that symmetrical vectors in Eq. (33.19) form a subspace of $\vec{\mathcal{H}}^{(c)}$.

Quantum Mechanics: Problems and Solutions
K. Kong Wan
Copyright © 2021 Jenny Stanford Publishing Pte. Ltd.
ISBN 978-981-4800-72-3 (Paperback), 978-0-429-29647-5 (eBook)
www.jennystanford.com

SQ33(2) The symmetrical vectors clearly form a linear subset since any finite linear combination combinations of symmetrical vectors are again symmetrical. To see that the set of symmetrical vectors form a subspace we need to check that any Cauchy sequence of symmetrical vectors would converge to a symmetrical vector.

Generally let $\vec{\phi}_\ell$ be a Cauchy sequence of vectors in $\vec{\mathcal{H}}^{(c)}$. Such a sequence would converge to a vector $\vec{\phi}$ in $\vec{\mathcal{H}}^{(c)}$. The permutation operator $\widehat{U}_p$ in Eq. (33.14) is a bounded operator. By Theorem 17.1(1) it is also a continuous operator. This means that the sequence $\widehat{U}_p\vec{\phi}_\ell$ would converge to $\widehat{U}_p\vec{\phi}$.

A Cauchy sequence of symmetrical vectors $\vec{\psi}_\ell^{(s)}$ would converge to a vector $\vec{\psi}$ in $\vec{\mathcal{H}}^{(c)}$. By definition the permutation operator does not affect symmetrical vectors, i.e., $\widehat{U}_p\vec{\psi}_\ell^{(s)} = \vec{\psi}_\ell^{(s)}$ as shown in Eq. (33.16). It follows that the sequence $\widehat{U}_P\vec{\psi}_\ell^{(s)}$ would also converge to $\vec{\psi}$. On the other hand the sequence $\widehat{U}_p\vec{\psi}_\ell^{(s)}$ must also converge to $\widehat{U}_p\vec{\psi}$ since $\widehat{U}_p$ is continuous. This means that the $\widehat{U}_p\vec{\psi} = \vec{\psi}$ which implies that the limit vector $\vec{\psi}$ is symmetrical. We can conclude that symmetrical vectors form a subspace of $\vec{\mathcal{H}}^{(c)}$.

We can also use the projector $\widehat{P}^{(cs)}$ in Eq. (33.22) to check the result. Since $\widehat{U}_p\widehat{P}^{(cs)} = \widehat{P}^{(cs)}$ we get $\widehat{U}_p\big(\widehat{P}^{(cs)}\vec{\phi}\,\big) = \widehat{P}^{(cs)}\vec{\phi}$, i.e., $\widehat{U}_p$ has no effect on the vector $\widehat{P}^{(cs)}\vec{\phi}$. This implies that the projector $\widehat{P}^{(cs)}$ projects every vector $\vec{\phi}$ in $\vec{\mathcal{H}}^{(c)}$ onto symmetrical vector. This means that the set of symmetrical vectors form the subspace onto which $\widehat{P}^{(cs)}$ projects.

Chapter 34

Conceptual Issues

Q34(1) Verify that $\vec{\Phi}^{(q,m)}(t)$ in Eq. (34.22) satisfies the initial condition in Eq. (34.16) and the Schrödinger equation (34.21).

SQ34(1) By setting $t = 0$ in Eq. (34.22) we can clearly see that $\vec{\Phi}^{(q,m)}(t)$ satisfies the initial condition in Eq. (34.16). To prove $\vec{\Phi}^{(q,m)}(t)$ satisfies the Schrödinger equation we first have

$$
\begin{aligned}
& i\hbar \frac{d}{dt} \vec{\Phi}^{(q,m)}(t) \\
&= i\rho c_- \vec{\varphi}_- \otimes \big(-\sin(\rho t/\hbar)\, \vec{\eta}^{(0)\oplus} - i \cos(\rho t/\hbar)\, \vec{\eta}^{(-)\oplus} \big) \\
&\quad + i\rho c_2\, \vec{\varphi}_+ \otimes \big(-\sin(\rho t/\hbar)\, \eta^{(0)\oplus} - i \cos(\rho t/\hbar)\, \vec{\eta}^{(+)\oplus} \big) \\
&= \rho c_- \vec{\varphi}_- \otimes \big(-i \sin(\rho t/\hbar)\, \vec{\eta}^{(0)\oplus} + \cos(\rho t/\hbar)\, \vec{\eta}^{(-)\oplus} \big) \\
&\quad + \rho c_2\, \vec{\varphi}_+ \otimes \big(-i \sin(\rho t/\hbar)\, \eta^{(0)\oplus} + \cos(\rho t/\hbar)\, \vec{\eta}^{(+)\oplus} \big).
\end{aligned}
$$

Next we have, using the Hamiltonian in Eq. (34.17)

$$
\begin{aligned}
& \widehat{H}^{(q,m)} \vec{\Phi}^{(q,m)}(t) \\
&= \rho \Big(\widehat{P}_-^{(q)} \otimes \widehat{L}_-^{(m)} + \widehat{P}_+^{(q)} \otimes \widehat{L}_+^{(m)} \Big) \vec{\Phi}^{(q,m)}(t) \\
&= \rho c_- \vec{\varphi}_- \otimes \big(\cos(\rho t/\hbar) \vec{\eta}^{(-)\oplus} - i\, \sin(\rho t/\hbar) \vec{\eta}^{(0)\oplus} \big) \\
&\quad + \rho c_+ \vec{\varphi}_+ \otimes \big(\cos(\rho t/\hbar) \eta^{(+)\oplus} - i\, \sin(\rho t/\hbar) \eta^{(0)\oplus} \big).
\end{aligned}
$$

The results above show that $\vec{\Phi}^{(q,m)}(t)$ satisfies the Schrödinger equation, i.e., it satisfies Eq. (34.21).

Quantum Mechanics: Problems and Solutions
K. Kong Wan

ISBN 978-981-4800-72-3 (Paperback), 978-0-429-29647-5 (eBook)
www.jennystanford.com

Chapter 35

Harmonic and Isotropic Oscillators

Q35(1) Verify the results in Eqs. (35.5) to (35.8).

SQ35(1) The probabilities are given by Postulate 28.1(PPDO), i.e., by Eq. (28.1). In other words we have

$$\wp^{H_{ho}}(\phi^s, E_1) = \langle \vec{\phi} \mid \widehat{P}^{\hat{H}_{ho}}(E_1)\vec{\phi}\,\rangle = \langle \vec{\phi} \mid |\vec{\varphi}_1\rangle\langle\vec{\varphi}_1|\vec{\phi}\,\rangle = 1/3,$$
$$\wp^{H_{ho}}(\phi^s, E_2) = \langle \vec{\phi} \mid \widehat{P}^{\hat{H}_{ho}}(E_2)\vec{\phi}\,\rangle = \langle \vec{\phi} \mid |\vec{\varphi}_2\rangle\langle\vec{\varphi}_2|\vec{\phi}\,\rangle = 2/3,$$
$$\wp^{H_{ho}}(\phi^s, E_n) = 0 \text{ if } n \neq 1, 2 \text{ since } |\vec{\varphi}_n\rangle\langle\vec{\varphi}_n|\,\vec{\phi} = \vec{0}.$$

Given that $E_1 = 3\hbar\omega/2$ and $E_2 = 5\hbar\omega/2$ we can calculate the expectation value $\mathcal{E}(H_{ho}, \phi^s)$, i.e.,

$$\mathcal{E}(H_{ho}, \phi^s) = \sum_{n=0}^{\infty} \wp^{H_{ho}}(\phi^s, E_n)\, E_n = \Big(\frac{1}{3}\frac{3}{2} + \frac{2}{3}\frac{5}{2}\Big)\hbar\omega = \frac{13}{6}\,\hbar\omega.$$

This is also is equal to $\langle \vec{\phi} \mid \widehat{H}_{ho}\vec{\phi}\,\rangle$ since for the state vector $\vec{\phi}$ in Eq. (35.4) we have

$$\begin{aligned}
\langle \vec{\phi} \mid \widehat{H}_{ho}\vec{\phi}\,\rangle &= \langle \frac{1}{\sqrt{3}}\,\vec{\varphi}_1 + \sqrt{\frac{2}{3}}\,\vec{\varphi}_2 \mid \widehat{H}_{ho}\Big(\frac{1}{\sqrt{3}}\,\vec{\varphi}_1 + \sqrt{\frac{2}{3}}\,\vec{\varphi}_2\Big)\rangle \\
&= \langle \frac{1}{\sqrt{3}}\,\vec{\varphi}_1 + \sqrt{\frac{2}{3}}\,\vec{\varphi}_2 \mid \Big(\frac{1}{\sqrt{3}}\,E_1\,\vec{\varphi}_1 + \sqrt{\frac{2}{3}}\,E_2\,\vec{\varphi}_2\Big)\rangle \\
&= \langle \frac{1}{\sqrt{3}}\,\vec{\varphi}_1 \mid E_1\frac{1}{\sqrt{3}}\,\vec{\varphi}_1\Big)\rangle + \langle \sqrt{\frac{2}{3}}\,\vec{\varphi}_2 \mid E_2\sqrt{\frac{2}{3}}\,\vec{\varphi}_2\,\rangle \\
&= \frac{1}{3}\,E_1 + \frac{2}{3}\,E_2 = \frac{1}{3}\frac{3}{2}\,\hbar\omega + \frac{2}{3}\frac{5}{2}\,\hbar\omega = \frac{13}{6}\,\hbar\omega.
\end{aligned}$$

Quantum Mechanics: Problems and Solutions
K. Kong Wan

ISBN 978-981-4800-72-3 (Paperback), 978-0-429-29647-5 (eBook)
www.jennystanford.com

Q35(2) Verify Eq. (35.10).

SQ35(2) Using Eq. (35.9) we get

$$[\widehat{a}, \widehat{a}^{\dagger}] = \frac{\lambda}{2}[\widehat{x} + \frac{i}{m\omega}\widehat{p}, \widehat{x} - \frac{i}{m\omega}\widehat{p}]$$
$$= \frac{\lambda}{2}\left(-\frac{i}{m\omega}[\widehat{x}, \widehat{p}] + \frac{i}{m\omega}[\widehat{p}, \widehat{x}]\right) = \widehat{I\!I},$$

using $[\widehat{x}, \widehat{p}] = i\hbar\widehat{I\!I}$ and $\lambda = m\omega/\hbar$. This is the result in Eq. (35.10), subject to domain restriction.

Q35(3) Verify that $\varphi_0(x)$ in Eq. (35.13) satisfies Eq. (35.12).

SQ35(3) First differentiate $\varphi_0(x)$ given in Eq. (35.13) to get

$$\frac{d}{dx}\varphi_0(x) = \left(\frac{\lambda}{\pi}\right)^{\frac{1}{4}}(-\lambda x)e^{-\frac{1}{2}\lambda x^2} = -\lambda x\,\varphi_0(x).$$

With $\lambda = m\omega/\hbar$ Eq. (35.12) follows immediately.

Q35(4) Obtain the first and the second excited state eigenvectors from the expression in Eq. (35.19) in terms of $\varphi_{H0}(x)$.

SQ35(4) The general expression of Hermite polynomials are given by Eqs. (16.14) and (16.15). Explicitly we have

$$H_0(y) = 1, \quad H_1(y) = 2y, \quad H_2(y) = 4y^2 - 2,$$

where $y = \sqrt{\lambda}\,x$. The corresponding Hermite functions defined by Eq. (16.18) are[1]

$$\varphi_{H0}(x) = \varphi_0(x) \text{ in Eq. (35.13)},$$
$$\varphi_{H1}(x) = \sqrt{2\lambda}\,x\varphi_0(x) = \sqrt{2\lambda}\,x\varphi_{H0},$$
$$\varphi_{H2}(x) = \frac{1}{\sqrt{2}}(2\lambda x^2 - 1)\,\varphi_0(x) = \frac{1}{\sqrt{2}}(2\lambda x^2 - 1)\,\varphi_{H0}(x).$$

Our problem is to check that Eq. (35.19) leads to the same results. Letting $n = 1$ and 2 in Eq. (35.17) we get[2]

$$\vec{\varphi}_{H1}(x) = \widehat{a}^{\dagger}\,\vec{\varphi}_{H0} := \sqrt{\frac{\lambda}{2}}\left(x - \frac{\hbar}{m\omega}\frac{d}{dx}\right)\varphi_{H0}(x)$$
$$= \sqrt{\frac{\lambda}{2}}\left(x\varphi_{H0}(x) - \frac{1}{\lambda}\frac{d}{dx}\varphi_{H0}(x)\right)$$
$$= \sqrt{2\lambda}\,x\varphi_{H0}(x) = \varphi_{H1}.$$

---

[1]Note that the factorial of zero is equal to 1, i.e., $0! = 1$.
[2]The results here can be compared with Eqs. (16.19) to (16.22).

$$\vec{\varphi}_{H2}(x) = \frac{1}{\sqrt{2}} \hat{a}^\dagger \vec{\varphi}_{H1}$$

$$:= \frac{\sqrt{\lambda}}{2} \left(x - \frac{\hbar}{m\omega} \frac{d}{dx} \right) \varphi_{H1}(x)$$

$$= \frac{\sqrt{\lambda}}{2} \left(x\varphi_{H1}(x) - \frac{1}{\lambda} \frac{d}{dx} \varphi_{H1}(x) \right)$$

$$= \frac{\sqrt{\lambda}}{2} \sqrt{2\lambda} \left(x^2 \varphi_{H0}(x) - \frac{1}{\lambda} \left(\varphi_{H0}(x) + x \frac{d}{dx} \varphi_{H0}(x) \right) \right)$$

$$= \frac{1}{\sqrt{2}} \lambda \left(x^2 \varphi_{H0}(x) - \frac{1}{\lambda} \varphi_{H0}(x) + x^2 \varphi_{H0}(x) \right)$$

$$= \frac{1}{\sqrt{2}} (2\lambda x^2 - 1) \varphi_{H0}(x) = \varphi_{H2}.$$

Q35(5) Verify Eq. (35.14) and (35.15).

SQ35(5) To verify Eq. (35.14) we have

$$\hat{a}^\dagger \hat{a} = \frac{\lambda}{2} \left(\hat{x} - \frac{i}{m\omega} \hat{p} \right) \left(\hat{x} + \frac{i}{m\omega} \hat{p} \right)$$

$$= \frac{\lambda}{2} \left(\hat{x}^2 + \frac{1}{(m\omega)^2} \hat{p}^2 + \frac{i}{m\omega} \hat{x}\hat{p} - \frac{i}{m\omega} \widehat{px} \right)$$

$$= \frac{\lambda}{2} \left(\hat{x}^2 + \frac{1}{(m\omega)^2} \hat{p}^2 + \frac{i}{m\omega} [\hat{x}, \hat{p}] \right)$$

$$= \frac{m\omega}{2\hbar} \left(\hat{x}^2 + \frac{1}{(m\omega)^2} \hat{p}^2 - \frac{\hbar}{m\omega} \hat{I\!I} \right).$$

To verify Eq. (35.15) we have

$$\hbar\omega \hat{a}^\dagger \hat{a} = \frac{1}{2m} \hat{p}^2 + \frac{1}{2} m\omega^2 \hat{x}^2 - \frac{1}{2} \hbar\omega$$

$$\Rightarrow \quad \hat{H}_{ho} = \hbar\omega \left(\hat{N} + \frac{1}{2} \right).$$

Q35(6) Verify that $\phi(x, t)$ in Eq. (35.39) is normalised and that it also satisfies the Schrödinger equation for time evolution of the harmonic oscillator.

SQ35(6) Using Eqs. (35.43) for $|\phi(x, t)|^2$ and following formula of integration

$$\int_{-\infty}^{\infty} e^{-a(x-b)^2} \, dx = \sqrt{\frac{\pi}{a}}$$

we can verify the normalisation of $\phi(x, t)$ immediately, i.e.,

$$\langle \vec{\phi}(t) \mid \vec{\phi}(t) \rangle = \int_{-\infty}^{\infty} |\phi(x, t)|^2$$
$$= \int_{-\infty}^{\infty} \left(\frac{\lambda}{\pi}\right)^{\frac{1}{2}} e^{-\lambda\left(x - x_c(t)\right)^2} dx = 1.$$

To verify the Schrödinger equation we can perform the following calculations. First let

$$f(x, t) = \exp\left(\frac{i}{\hbar} p_c(t)x - \frac{m\omega}{2\hbar}\left(x - x_c(t)\right)^2\right).$$

Then we have $\phi(x, t) = C(t) f(x, t)$. We can carry out the following calculations:

(1) For the time derivative we have

$$\frac{dx_c(t)}{dt} = -\omega\, x_c(0) \sin \omega t = \frac{1}{m} p_c(t),$$
$$\frac{dp_c(t)}{dt} = -m\omega^2 x_c(0) \cos \omega t = -m\omega^2 x_c(t),$$
$$i\hbar\frac{dC(t)}{dt} = \left(\frac{1}{2m} p_c^2(t) - \frac{1}{2} m\omega^2 x_c^2(t) + \frac{1}{2}\hbar\omega\right) C(t),$$
$$i\hbar\frac{\partial f(x, t)}{\partial t} = \left(m\omega^2 x_c(t)x + i\omega\left(x - x_c(t)\right) p_c(t)\right) f(x, t),$$
$$i\hbar\frac{\partial \phi(x, t)}{\partial t} = i\hbar\frac{dC(t)}{dt} f(x, t) + C(t)\left(i\hbar\frac{\partial f(x, t)}{\partial t}\right)$$
$$= \left(\frac{1}{2m} p_c^2(t) - \frac{1}{2} m\omega^2 x_c^2(t) + \frac{1}{2}\hbar\omega\right) \phi(x, t)$$
$$+ \left(m\omega^2 x_c(t)\, x + i\omega(x - x_c(t))\, p_c(t)\right)\phi(x, t).$$

(2) For the spatial derivatives derivative we have

$$\widehat{p}\,\phi(x, t) = \left(p_c(t) + im\omega\left(x - x_c(t)\right)\right)\phi(x, t).$$
$$\widehat{p}^2\phi(x, t) = \left(m\hbar\omega + \left(p_c(t) + im\omega\left(x - x_c(t)\right)\right)^2\right)\phi(x, t).$$
$$\frac{\widehat{p}^2}{2m}\phi(x, t) = \left(\frac{1}{2}\hbar\omega + \frac{1}{2m} p_c^2(t) + i\omega\, p_c(t)(x - x_c(t))\right.$$
$$\left. - \frac{1}{2} m\omega^2 x^2 + m\omega^2\, x\, x_c(t) - \frac{1}{2} m\omega^2 x_c^2(t)\right)\phi(x, t).$$

(3) Comparing the above results we get

$$i\hbar\frac{\partial \phi(x, t)}{\partial t} = \widehat{H}_{ho}\phi(x, t).$$

In other words the wave function $\phi(x, t)$ satisfies the Schrödinger equation.

Q35(7) Verify Eqs. (35.44), (35.45).

SQ35(7) Using Eqs. (35.23) and (35.27) for $\phi(x, t)$ and $|\phi(x, t)|^2$ we get

$$\langle \vec{\phi}(t) \mid \hat{x}\,\vec{\phi}(t)\rangle = \int_{-\infty}^{\infty} x\,|\phi(x, t)|^2$$

$$= \int_{-\infty}^{\infty} \left(\frac{\lambda}{\pi}\right)^{\frac{1}{2}} x\, e^{-\lambda\left(x - x_c(t)\right)^2} dx.$$

We can rewrite the integral as the sum of two integrals, i.e.,

$$\int_{-\infty}^{\infty} \left(\frac{\lambda}{\pi}\right)^{\frac{1}{2}} \left(x - x_c(t)\right) e^{-\lambda\left(x - x_c(t)\right)^2} dx$$

$$+ \int_{-\infty}^{\infty} \left(\frac{\lambda}{\pi}\right)^{\frac{1}{2}} x_c(t)\, e^{-\frac{1}{2}\lambda\left(x - x_c(t)\right)^2} dx.$$

The first integral can be written as

$$\int_{-\infty}^{\infty} \left(\frac{\lambda}{\pi}\right)^{\frac{1}{2}} y e^{-\lambda y^2} dy, \quad y = x - x_c(t).$$

This integral vanishes on account of the integrand being an odd function of y. Since the function $\phi(x, t)$ is normalised the second integral is equal to $x_c(t)$, i.e., we get

$$\langle \vec{\phi}(t) \mid \hat{x}\,\vec{\phi}(t)\rangle = x_c(t).$$

To calculate $\langle \vec{\phi}(t) \mid \hat{p}\,\vec{\phi}(t)\rangle$ we first observe that

$$\hat{p}\,\phi(x, t) = \Big(p_c(t) + im\omega\big(x - x_c(t)\big)\Big)\,\phi(x, t).$$

The integral for $\langle \vec{\phi}(t) \mid \hat{p}\,\vec{\phi}(t)\rangle$ becomes

$$\int_{-\infty}^{\infty} \Big(p_c(t) + im\omega\big(x - x_c(t)\big)\Big) |\,\phi(x, t)|^2\, dx$$

$$= p_c(t) \int_{-\infty}^{\infty} |\,\phi(x, t)|^2\, dx + \int_{-\infty}^{\infty} \big(-im\omega(x - x_c(t)\big)|\,\phi(x, t)|^2\, dx$$

$$= p_c(t),$$

since the second integral vanishes due to an odd integrand.

Q35(8) Verify Eq. (35.46).

SQ35(8) First we evaluate $\langle \vec{\phi}(t) \mid \widehat{x}^2\, \vec{\phi}(t)\rangle$ which is given by

$$\langle \vec{\phi}(t) \mid \widehat{x}^2\, \vec{\phi}(t)\rangle = \int_{-\infty}^{\infty} \left(\frac{\lambda}{\pi}\right)^{\frac{1}{2}} x^2\, e^{-\lambda\left(x - x_c(t)\right)^2}\, dx. \qquad (*)$$

In order to employ the formula

$$\int_{-\infty}^{\infty} x^2 e^{-ax^2}\, dx = \frac{1}{2a}\sqrt{\frac{\pi}{a}}, \qquad (**)$$

we replace x^2 in the integrand of the integral in Eq. (∗) above by

$$\left(x - x_c(t) + x_c(t)\right)^2 = \left(x - x_c(t)\right)^2 + 2(x - x_c(t))x_c(t) + x_c(t)^2.$$

The integral in Eq. (∗) becomes a sum of three integrals, i.e.,

$$\int_{-\infty}^{\infty} \left(\frac{\lambda}{\pi}\right)^{\frac{1}{2}} \left(x - x_c(t)\right)^2 \exp\left(-\lambda\left(x - x_c(t)\right)^2\right) dx$$

$$+ \int_{-\infty}^{\infty} \left(\frac{\lambda}{\pi}\right)^{\frac{1}{2}} 2\left(x - x_c(t)\right) x_c(t) \exp\left(-\lambda\left(x - x_c(t)\right)^2\right) dx$$

$$+ \int_{-\infty}^{\infty} \left(\frac{\lambda}{\pi}\right)^{\frac{1}{2}} x_c(t)^2 \exp\left(-\lambda\left(x - x_c(t)\right)^2\right) dx.$$

The values of these integrals are as follows:

(1) On account of the formula in Eq. (∗∗) above for the first integral can be evaluated to give the value $1/2\lambda = \hbar/2m\omega$.
(2) The second integral vanishes due to an odd integrand.
(3) The third integral has a value of $x_c(t)^2$ due to $\phi(x, t)$ being normalised.

The final result is

$$\langle \vec{\phi}_t \mid \widehat{x}^2\, \vec{\phi}_t\rangle = x_c(t)^2 + \frac{\hbar}{2m\omega}.$$

Next we evaluate $\langle \vec{\phi}(t) \mid \widehat{p}^2\, \vec{\phi}(t)\rangle$ which is given by

$$\langle \vec{\phi}(t) \mid \widehat{p}^2\, \vec{\phi}(t)\rangle = \int_{-\infty}^{\infty} \phi^*(x, t)\, \widehat{p}^2 \phi(x, t) dx.$$

Using the result obtained for SQ35(6) for $\widehat{p}^2\phi(x, t)$ we can see that $\langle \vec{\phi}(t) \mid \widehat{p}^2\, \vec{\phi}(t)\rangle$ is given by the following integral:

$$\int_{-\infty}^{\infty} \left(\hbar m\omega + \left(p_c(t) + i m\omega\left(x - x_c(t)\right)\right)^2\right) |\phi(x, t)|^2 dx.$$

Making used of previous results this integral can be evaluated as follows:

(1) The integral over $\hbar m\omega$ has the value $\hbar m\omega$.
(2) The integral over $p_c(t)^2$ has the value $p_c(t)^2$.
(3) The integral over $2i\,p_c(t)m\omega\big(x - x_c(t)\big)$ has the value 0 due to an odd integrand.
(4) The integral over $-m^2\omega^2\big(x - x_c(t)\big)^2$ has the value $-\hbar m\omega/2$.

The final result is

$$\langle \vec{\phi}(t) \mid \widehat{p}^{\,2}\, \vec{\phi}(t)\rangle = p_c(t)^2 + \frac{1}{2}\, m\hbar\omega.$$

It follows that the expectation value of the Hamiltonian is given by

$$\begin{aligned}\langle \vec{\phi}(t) \mid \widehat{H}\, \vec{\phi}(t)\rangle &= \frac{1}{2m}\,\langle \vec{\phi}(t) \mid \widehat{p}^{\,2}\, \vec{\phi}(t)\rangle + \frac{1}{2}\, m\omega^2 \langle \vec{\phi}(t) \mid \widehat{x}^{2}\, \vec{\phi}(t)\rangle \\ &= \frac{1}{2m}\, p_c(t)^2 + \frac{1}{2}\, m\omega^2 x_c(t)^2 + \frac{1}{2}\,\hbar\omega.\end{aligned}$$

This differs from the corresponding classical value by $\hbar\omega/2$.

Q35(9) Using the expressions in Eqs. (35.40) and (35.41) for $x_c(t)$ and $p_c(t)$ show that energy expectation value in Eq. (35.46) is conserved, i.e., the value is time-independent. Explain how the expectation value can approximate the energy of a classical harmonic oscillator.

SQ35(9) The explicit time-dependence of the energy expectation value in Eq. (35.46) is given as follows:

$$\begin{aligned}&\langle \vec{\phi}(t) \mid \widehat{H}\, \vec{\phi}(t)\rangle \\ &= \frac{1}{2m}\, p_c^2(t) + \frac{1}{2}\, m\omega^2\, x_c^2(t) + \frac{1}{2}\,\hbar\omega \\ &= \frac{1}{2m}\big(-m\omega x_c(0)\sin\omega t\big)^2 + \frac{1}{2}\, m\omega^2\, x_c^2(0)\cos^2\omega t + \frac{1}{2}\,\hbar\omega \\ &= \frac{m\omega^2}{2}\, x_c^2(0)\big(\cos^2\omega t + \sin^2\omega t\big) + \frac{1}{2}\,\hbar\omega \\ &= \frac{m\omega^2}{2}\, x_c^2(0) + \frac{1}{2}\,\hbar\omega,\end{aligned}$$

which is time-independent, i.e., the energy expectation value $\mathcal{E}\big(\widehat{H}_{ho}, \vec{\phi}\big)$ is conserved.

The first term in the above expression for the expectation value is equal to the classical value for a classical oscillator initially placed at rest at position $x_c(0)$. For sufficient large oscillations due to a large initial displacement $x_c(0)$ the second term $\hbar\omega/2$ becomes negligible. The above expectation value then approximates the energy of a corresponding classical harmonic oscillator.

Q35(10) Verify Eqs. (35.58) and (35.59).

SQ35(10) Using Eqs. (35.56) and (35.57) we have:

(1) For $\langle\vec{\phi}(0) \mid \widehat{x}_{Hei}(t)\vec{\phi}(0)\rangle$ we have

$$\begin{aligned}&\langle\vec{\phi}(0) \mid \widehat{x}_{Hei}(t)\vec{\phi}(0)\rangle\\ &= (\cos\omega t)\,\langle\vec{\phi}(0) \mid \widehat{x}\,\vec{\phi}(0)\rangle + \left(\frac{\sin\omega t}{m\omega}\right)\langle\vec{\phi}(0) \mid \widehat{p}\,\vec{\phi}(0)\rangle\\ &= (\cos\omega t)\, x_c(0) = x_c(t).\end{aligned}$$

We have used Eqs. (35.44) and (35.55) at $t = 0$ proved in SQ35(7), i.e., $\langle\vec{\phi}(0) \mid \widehat{x}\,\vec{\phi}(0)\rangle = x_c(0)$ and $\langle\vec{\phi}(0) \mid \widehat{p}\,\vec{\phi}(0)\rangle = p_c(0) = 0$.

(2) For $\langle\vec{\phi}(0) \mid \widehat{p}_{Hei}(t)\vec{\phi}(0)\rangle$ we have

$$\begin{aligned}&\langle\vec{\phi}(0) \mid \widehat{p}_{Hei}(t)\vec{\phi}(0)\rangle\\ &= (-m\omega\sin\omega t)\,\langle\vec{\phi}(0) \mid \widehat{x}\,\vec{\phi}(0)\rangle + (\cos\omega t)\,\langle\vec{\phi}(0) \mid \widehat{p}\,\vec{\phi}(0)\rangle\\ &= (-m\omega\sin\omega t)\, x_c(0) = p_c(t).\end{aligned}$$

Q35(11) Verify Eq. (35.60) by explicit calculation of $\widehat{H}_{hoHei}(t)$ using $\widehat{x}_{Hei}(t)$ and $\widehat{p}_{Hei}(t)$ in Eqs. (35.56) and (35.57).

SQ35(11) From Eq. (35.47) we have

$$\begin{aligned}\widehat{H}_{hoHei}(t) &= \frac{1}{2m}\,\widehat{p}^{\,2}_{Hei}(t) + \frac{1}{2}\,m\omega^2\widehat{x}^{\,2}_{Hei}(t)\\ &= \frac{1}{2m}\Big((-m\omega\sin\omega t)\,\widehat{x} + (\cos\omega t)\,\widehat{p}\Big)^2\\ &\quad + \frac{1}{2}\,m\omega^2\left((\cos\omega t)\,\widehat{x} + \left(\frac{\sin\omega t}{m\omega}\right)\widehat{p}\right)^2\\ &= \frac{1}{2m}\,\widehat{p}^{\,2} + \frac{1}{2}\,m\omega^2\,\widehat{x}^{\,2} = \widehat{H}_{hoHei}(0).\end{aligned}$$

We have used the following results:

$$\begin{aligned}\Big((-m\omega\sin\omega t)\,\widehat{x} + (\cos\omega t)\,\widehat{p}\Big)^2 &= (-m\omega\sin\omega t)^2\,\widehat{x}^{\,2} + (\cos\omega t)^2\,\widehat{p}^{\,2}\\ &\quad - m\omega(\sin\omega t)(\cos\omega t)(\widehat{x}\,\widehat{p} + \widehat{p}\,\widehat{x}),\end{aligned}$$

$$\left((\cos\omega t)\,\widehat{x} + \left(\frac{\sin\omega t}{m\omega}\right)\widehat{p}\right)^2 = (\cos\omega t)^2\,\widehat{x}^2 + \left(\frac{\sin\omega t}{m\omega}\right)^2\widehat{p}^2 + \frac{(\sin\omega t)(\cos\omega t)}{m\omega}(\widehat{x}\,\widehat{p} + \widehat{p}\,\widehat{x}),$$

$$(\sin\omega t)^2 + (\cos\omega t)^2 = 1.$$

Q35(12) The Hamiltonian of a forced harmonic oscillator in the Schrödinger picture is given in the usual notation by[3]

$$\widehat{H}_{Sch} = \frac{1}{2m}\,\widehat{p}^{\,2}_{Sch} + \frac{1}{2}\,m\omega^2\,\widehat{x}^{\,2}_{Sch} - g\left(\widehat{x}_{Sch} + \frac{1}{m\omega}\,\widehat{p}_{Sch}\right),$$

where g is a real number. Let $\widehat{a}_{Sch}$ and $\widehat{a}^{\dagger}_{Sch}$ be a pair of operators related to $\widehat{x}_{Sch}$ and $\widehat{p}_{Sch}$ by Eq. (35.9).

(a) Show that

$$\widehat{H}_{Sch} = \hbar\omega\left(\widehat{a}^{\dagger}_{Sch}\,\widehat{a}_{Sch} + \frac{1}{2}\right) + \gamma\,\widehat{a}_{Sch} + \gamma^*\,\widehat{a}^{\dagger}_{Sch},$$

where

$$\gamma = -\sqrt{\frac{\hbar}{2m\omega}}\,(1-i)g.$$

(b) In the Heisenberg picture the annihilation and creation operators become $\widehat{a}_{Hei}(t)$ and $\widehat{a}^{\dagger}_{Hei}(t)$ and the Hamiltonian becomes

$$\widehat{H}_{Hei}(t) = \hbar\omega\left(\widehat{a}^{\dagger}_{Hei}(t)\widehat{a}_{Hei}(t) + \frac{1}{2}\right) + \gamma\,\widehat{a}_{Hei}(t) + \gamma^*\,\widehat{a}^{\dagger}_{Hei}(t).$$

Show that $\widehat{a}_{Hei}(t)$ satisfies the Heisenberg equation of motion and that the Heisenberg equation can be wrtten as

$$\frac{d}{dt}\widehat{a}_{Hei}(t) = -i\omega\widehat{a}_{Hei}(t) - \frac{i}{\hbar}\gamma^*.$$

Show further that this equation can be rewritten in the form

$$\frac{d}{dt}\left(\widehat{a}_{Hei}(t)e^{i\omega t}\right) = -\frac{i}{\hbar}\gamma^*\,e^{i\omega t}.$$

Integrate this equation to obtain an explicit expression for the time dependence of $\widehat{a}_{Hei}(t)$.

[3] Merzbacher pp. 335–336.

SQ35(12)(a) In the Schrödinger picture we have

$$\widehat{a}_{Sch} := \sqrt{\frac{\lambda}{2}}\Big(\widehat{x}_{Sch} + \frac{i}{m\omega}\,\widehat{p}_{Sch}\Big),$$
$$\widehat{a}^{\dagger}_{Sch} := \sqrt{\frac{\lambda}{2}}\Big(\widehat{x}_{Sch} - \frac{i}{m\omega}\,\widehat{p}_{Sch}\Big).$$

We can calculate the product $\widehat{a}^{\dagger}_{Sch}\widehat{a}_{Sch}$ in terms of $\widehat{x}_{Sch}$ and $\widehat{p}_{Sch}$ as in SQ35(5), i.e.,

$$\begin{aligned}
\widehat{a}^{\dagger}_{Sch}\widehat{a}_{Sch} &= \frac{m\omega}{2\hbar}\left(\widehat{x}_{Sch} - \frac{i}{m\omega}\widehat{p}_{Sch}\right)\left(\widehat{x}_{Sch} + \frac{i}{m\omega}\widehat{p}_{Sch}\right) \\
&= \frac{m\omega}{2\hbar}\left(\widehat{x}^{2}_{Sch} + \frac{1}{(m\omega)^2}\widehat{p}^{2}_{Sch} + \frac{i}{m\omega}(\widehat{x}_{Sch}\widehat{p}_{Sch} - \widehat{p}_{Sch}\widehat{x}_{Sch})\right) \\
&= \frac{m\omega}{2\hbar}\,\widehat{x}^{2}_{Sch} + \frac{1}{2m\hbar\omega}\,\widehat{p}^{2}_{Sch} - \frac{1}{2}.
\end{aligned}$$

It follows that

$$\hbar\omega\,\widehat{a}^{\dagger}_{Sch}\widehat{a}_{Sch} = \frac{1}{2m}\,\widehat{p}^{2}_{Sch} + \frac{1}{2}m\omega^2\,\widehat{x}^{2}_{Sch} - \frac{1}{2}\,\hbar\omega.$$

We can express $\widehat{x}_{Sch}$ and $\widehat{p}_{Sch}$ in terms of the annihilation and creation operators, i.e.,

$$\widehat{x}_{Sch} = \sqrt{\frac{\hbar}{2m\omega}}\,(\widehat{a}_{Sch} + \widehat{a}^{\dagger}_{Sch}),$$
$$\widehat{p}_{Sch} = \left(-i\sqrt{\frac{\hbar m\omega}{2}}\,(\widehat{a}_{Sch} - \widehat{a}^{\dagger}_{Sch})\right).$$

It follows that

$$\begin{aligned}
\widehat{x}_{Sch} + \frac{1}{m\omega}\,\widehat{p}_{Sch} &= \sqrt{\frac{\hbar}{2m\omega}}\,(\widehat{a}_{Sch} + \widehat{a}^{\dagger}_{Sch}) + \frac{1}{m\omega}\left(-i\sqrt{\frac{\hbar m\omega}{2}}\,(\widehat{a}_{Sch} - \widehat{a}^{\dagger}_{Sch})\right) \\
&= \sqrt{\frac{\hbar}{2m\omega}}\big(1-i\big)\,\widehat{a}_{Sch} + \sqrt{\frac{\hbar}{2m\omega}}\big(1+i\big)\,\widehat{a}^{\dagger}_{Sch} \\
\Rightarrow\quad & -g\,(\widehat{x}_{Sch} + \frac{1}{m\omega}\,\widehat{p}_{Sch}) \\
&= -g\sqrt{\frac{\hbar}{2m\omega}}\big(1-i\big)\,\widehat{a}_{Sch} - g\sqrt{\frac{\hbar}{2m\omega}}\big(1+i\big)\,\widehat{a}^{\dagger}_{Sch} \\
&= \gamma\,\widehat{a}_{Sch} + \gamma^*\,\widehat{a}^{\dagger}_{Sch} \\
\Rightarrow \widehat{H}_{Sch} &= \hbar\omega\Big(\widehat{a}^{\dagger}_{Sch}\widehat{a}_{Sch} + \frac{1}{2}\Big) + \gamma\widehat{a}_{Sch} + \gamma^*\widehat{a}^{\dagger}_{Sch}.
\end{aligned}$$

SQ35(12)(b) Not being selfadjoint the annihilation and creation operators do not represent any observables. Postulate 29.3.1(HP) does not directly apply to them. However, they are related to position and momentum at time $t = 0$ by Eqs. (35.9), i.e.,

$$\widehat{a} = \sqrt{\frac{\lambda}{2}}\left(\widehat{x} + \frac{i}{m\omega}\widehat{p}\right), \quad \widehat{a}^\dagger = \sqrt{\frac{\lambda}{2}}\left(\widehat{x} - \frac{i}{m\omega}\widehat{p}\right).$$

These annihilation and creation operators are identifiable with $\widehat{a}_{Sch}$ and $\widehat{a}^\dagger_{Sch}$ in SQ35(12)(a). In the Heisenberg picture the position and momentum operators will evolve into $\widehat{x}_{Hei}(t)$ and $\widehat{p}_{Hei}(t)$. It follows that the annihilation and creation operators will evolve into

$$\widehat{a}_{Hei}(t) = \sqrt{\frac{\lambda}{2}}\left(\widehat{x}_{Hei}(t) + \frac{i}{m\omega}\widehat{p}_{Hei}(t)\right),$$

$$\widehat{a}^\dagger_{Hei}(t) = \sqrt{\frac{\lambda}{2}}\left(\widehat{x}_{Hei}(t) - \frac{i}{m\omega}\widehat{p}_{Hei}(t)\right),$$

and that the Hamiltonian $\widehat{H}_{Sch}$ will evolve into

$$\widehat{H}_{Hei}(t) = \hbar\omega\widehat{a}^\dagger_{Hei}(t)\widehat{a}_{Hei}(t) + \frac{1}{2}\hbar\omega + \gamma\,\widehat{a}_{Hei}(t) + \gamma^*\,\widehat{a}^\dagger_{Hei}(t).$$

These time-dependent operators possess the following properties:

(1) Using the commutation relation between $\widehat{x}_{Hei}(t)$ and $\widehat{p}_{Hei}(t)$, i.e., $[\widehat{x}_{Hei}(t), \widehat{p}_{Hei}(t)] = i\hbar\widehat{I\!I}$, we can verify that the commutation relation between the annihilation and creation operators are preserved, i.e., we have, from Eq. (35.10),

$$[\widehat{a}_{Hei}(t), \widehat{a}^\dagger_{Hei}(t)] = \widehat{I\!I}.$$

(2) $\widehat{a}_{Hei}(t)$ satisfies the Heisenberg equation of motion since

$$\begin{aligned} i\hbar\frac{d}{dt}\widehat{a}_{Hei}(t) &= \sqrt{\frac{\lambda}{2}}\left(i\hbar\frac{d}{dt}\widehat{x}_{Hei}(t) + \frac{i}{m\omega}i\hbar\frac{d}{dt}\widehat{p}_{Hei}(t)\right) \\ &= \sqrt{\frac{\lambda}{2}}\left([\widehat{x}_{Hei}(t), \widehat{H}_{Hei}(t)] + \frac{i}{m\omega}[\widehat{p}_{Hei}(t), \widehat{H}_{Hei}(t)]\right) \\ &= [\widehat{a}_{Hei}(t), \widehat{H}_{Hei}(t)]. \end{aligned}$$

The above commutator can be calculated, i.e.,

$$\begin{aligned} [\widehat{a}_{Hei}(t), \widehat{H}_{Hei}] &= [\widehat{a}_{Hei}(t), \hbar\omega\left(\widehat{a}^\dagger_{Hei}(t)\widehat{a}_{Hei}(t) + \frac{1}{2}\right) \\ &\quad + \gamma\,\widehat{a}_{Hei}(t) + \gamma^*\,\widehat{a}^\dagger_{Hei}(t)] \\ &= \hbar\omega\,\widehat{a}_{Hei}(t) + \gamma^*. \end{aligned}$$

The Heisenberg equation becomes

$$i\hbar\frac{d}{dt}\hat{a}_{Hei}(t) = \hbar\omega\hat{a}_{Hei}(t) + \gamma^*, \quad \text{or}$$

$$\frac{d}{dt}\hat{a}_{Hei}(t) = -i\omega\hat{a}_{Hei}(t) - \frac{i}{\hbar}\gamma^*.$$

It follows that:

$$\frac{d}{dt}\left(\hat{a}_{Hei}(t)e^{i\omega t}\right) = \left(\frac{d}{dt}\hat{a}_{Hei}(t)\right) e^{i\omega t} + i\omega\hat{a}_{Hei}(t)e^{i\omega t}$$

$$= \left(-i\omega\hat{a}_{Hei}(t) - \frac{i}{\hbar}\gamma^*\right) e^{i\omega t} + i\omega\hat{a}_{Hei}(t)e^{i\omega t}$$

$$= -\frac{i}{\hbar}\gamma^* e^{i\omega t}.$$

Integrating the above equation, i.e.,

$$\frac{d}{dt}\left(\hat{a}_{Hei}(t)e^{i\omega t}\right) = -\frac{i}{\hbar}\gamma^* e^{i\omega t},$$

from 0 to t we get

$$\hat{a}_{Hei}(t)e^{i\omega t} - \hat{a}_{Hei}(0) = -\frac{1}{\hbar\omega}\gamma^* e^{i\omega t} + \frac{1}{\hbar\omega}\gamma^*$$

$$\Rightarrow \quad \hat{a}_{Hei}(t) = \hat{a}_{Hei}(0)e^{-i\omega t} - \frac{1}{\hbar\omega}\gamma^* \left(1 - e^{-i\omega t}\right).$$

Q35(13) What is the degeneracy of the eigenvalues $E_{1,1}$, $E_{2,2}$ of the two-dimensional isotropic oscillator? Are all eigenvalues degenerate?

SQ35(13) From Eq. (35.86) we can see that the eigenvalue $E_{1,1}$ of a two-dimensional isotropic oscillator is degenerate with a degeneracy of 3 corresponding to the eigenvectors $\vec{\varphi}_{1,1}$, $\vec{\varphi}_{2,0}$ and $\vec{\varphi}_{0,2}$. The eigenvalue $E_{2,2}$ is degenerate with a degeneracy of 5 corresponding to the eigenvectors $\vec{\varphi}_{2,2}$, $\vec{\varphi}_{4,0}$, $\vec{\varphi}_{1,3}$, $\vec{\varphi}_{3,1}$ and $\vec{\varphi}_{0,4}$. Not all eigenvalues are degenerate. The ground state energy eigenvalue $E_{0,0}$ is nondegenerate with corresponding eigenvector $\vec{\varphi}_{0,0}$.

Chapter 36

Angular Momenta

Q36(1) The spherical harmonics $Y_{1,1}(\theta, \varphi)$, $Y_{1,0}(\theta, \varphi)$, $Y_{1,-1}(\theta, \varphi)$ are given by Eqs. (16.65) to (16.67). In Cartesian coordinates, these functions are given by[1]

$$Y_{1,-1}(x, y, z) = \sqrt{\frac{3}{8\pi}}\,\frac{x - iy}{r},$$

$$Y_{1,0}(x, y, z) = \sqrt{\frac{3}{4\pi}}\,\frac{z}{r},$$

$$Y_{1,1}(x, y, z) = -\sqrt{\frac{3}{8\pi}}\,\frac{x + iy}{r}.$$

(a) Using the expression in Cartesian coordinates for the quantised orbital angular momentum operator $\widehat{L}_{cz}$ in Eq. (27.86), verify by explicit calculations that $Y_{1,1}(x, y, z)$, $Y_{1,0}(x, y, z)$ and $Y_{1,-1}(x, y, z)$ are the eigenfunctions of $\widehat{L}_{cz}$ corresponding to eigenvalues $\hbar$, 0 and $-\hbar$.

(b) Consider the following coordinate transformations:

$$x \to z', \quad y \to x', \quad z \to y'.$$

[1]Zettili pp. 293–934. For convenience we have used the same symbols, e.g., $Y_{1,0}$ for the functions in Cartesian coordinates.

Quantum Mechanics: Problems and Solutions
K. Kong Wan

ISBN 978-981-4800-72-3 (Paperback), 978-0-429-29647-5 (eBook)
www.jennystanford.com

(1) Show that the component of the orbital angular momentum operator along the z'-direction is the same as that along the x-direction, i.e., show that $\widehat{L}_{cz'} = \widehat{L}_{cz}$.

(2) Show that the simultaneous eigenfunctions of $\widehat{L}_c^2$ and $\widehat{L}_{cx}$ corresponding to eigenvalues of $\widehat{L}_c^2$ equal to $2\hbar^2$ are given in Cartesian coordinates by

$$X_{1,-1}(x,y,z) = \sqrt{\frac{3}{8\pi}}\,\frac{y-iz}{r},$$

$$X_{1,0}(x,y,z) = \sqrt{\frac{3}{4\pi}}\,\frac{x}{r},$$

$$X_{1,1}(x,y,z) = -\sqrt{\frac{3}{8\pi}}\,\frac{y+iz}{r}.$$

SQ36(1)(a) In Cartesian coordinates we have[2]:

$$\widehat{p}_x \vec{Y}_{1,-1} := -i\hbar\sqrt{\frac{3}{8\pi}}\,\frac{\partial}{\partial x}\,\frac{x-iy}{r}$$

$$= -i\hbar\sqrt{\frac{3}{8\pi}}\left(\frac{1}{r} - \frac{x-iy}{r^2}\,\frac{\partial}{\partial x}\,r\right)$$

$$= -i\hbar\sqrt{\frac{3}{8\pi}}\left(\frac{1}{r} - \frac{x-iy}{r^2}\,\frac{x}{r}\right)$$

$$= -i\hbar\sqrt{\frac{3}{8\pi}}\,\frac{1}{r^3}\left(r^2 - x^2 + ixy\right).$$

$$\widehat{p}_y \vec{Y}_{1,-1} := -i\hbar\sqrt{\frac{3}{8\pi}}\,\frac{\partial}{\partial y}\,\frac{x-iy}{r}$$

$$= -i\hbar\sqrt{\frac{3}{8\pi}}\left(\frac{-i}{r} - \frac{x-iy}{r^2}\,\frac{\partial}{\partial y}\,r\right)$$

$$= -i\hbar\sqrt{\frac{3}{8\pi}}\left(\frac{-i}{r} - \frac{x-iy}{r^2}\,\frac{y}{r}\right)$$

$$= -i\hbar\sqrt{\frac{3}{8\pi}}\,\frac{1}{r^3}\left(-ir^2 - xy + iy^2\right).$$

[2] We have omitted the subscript c in the operators, e.g., writing $\widehat{p}_{cx}$ as $\widehat{p}_x$ in accordance with Eq. (27.90).

$$\begin{aligned}\widehat{L}_z\vec{Y}_{1,-1} &:= \left(x\widehat{p}_y - y\widehat{p}_x\right)Y_{1,-1}\\ &= -i\hbar\sqrt{\frac{3}{8\pi}}\frac{1}{r^3}\Big(\left(-ixr^2 - x^2y + ixy^2\right)\\ &\quad -\left(yr^2 - yx^2 + ixy^2\right)\Big)\\ &= -i\hbar\sqrt{\frac{3}{8\pi}}\frac{1}{r^3}\left(-ixr^2 - yr^2\right)\\ &= -\hbar\sqrt{\frac{3}{8\pi}}\frac{1}{r}\{x - iy\} = -\hbar\,Y_{1,-1}.\end{aligned}$$

Next we have

$$\widehat{p}_x\vec{Y}_{1,0} := -i\hbar\sqrt{\frac{3}{4\pi}}\frac{\partial}{\partial x}\frac{z}{r} = -i\hbar\sqrt{\frac{3}{4\pi}}\left(-\frac{z}{r^2}\frac{\partial}{\partial x}r\right) = i\hbar\sqrt{\frac{3}{4\pi}}\frac{zx}{r^3}.$$

$$\begin{aligned}\widehat{p}_y\vec{Y}_{1,0} &:= -i\hbar\sqrt{\frac{3}{4\pi}}\frac{\partial}{\partial y}\frac{z}{r} = -i\hbar\sqrt{\frac{3}{4\pi}}\left(-\frac{z}{r^2}\frac{\partial}{\partial y}r\right)\\ &= i\hbar\sqrt{\frac{3}{4\pi}}\frac{zy}{r^3}.\end{aligned}$$

$$\begin{aligned}\widehat{L}_z\vec{Y}_{1,0} &:= \left(x\widehat{p}_y - y\widehat{p}_x\right)Y_{1,0}\\ &= x\left(i\hbar\sqrt{\frac{3}{4\pi}}\frac{zy}{r^3}\right) - y\left(i\hbar\sqrt{\frac{3}{4\pi}}\frac{zx}{r^3}\right) = 0.\end{aligned}$$

Finally we have

$$\begin{aligned}\widehat{p}_x\vec{Y}_{1,1} &:= i\hbar\sqrt{\frac{3}{8\pi}}\frac{\partial}{\partial x}\frac{x+iy}{r} = -i\hbar\sqrt{\frac{3}{8\pi}}\left(\frac{1}{r} - \frac{x+iy}{r^2}\frac{\partial}{\partial x}r\right)\\ &= i\hbar\sqrt{\frac{3}{8\pi}}\left(\frac{1}{r} - \frac{x+iy}{r^2}\frac{x}{r}\right)\\ &= i\hbar\sqrt{\frac{3}{8\pi}}\frac{1}{r^3}\left(r^2 - x^2 - ixy\right),\end{aligned}$$

$$\begin{aligned}\widehat{p}_y\vec{Y}_{1,1} &:= i\hbar\sqrt{\frac{3}{8\pi}}\frac{\partial}{\partial y}\frac{x+iy}{r} = i\hbar\sqrt{\frac{3}{8\pi}}\left(\frac{i}{r} - \frac{x+iy}{r^2}\frac{\partial}{\partial y}r\right)\\ &= i\hbar\sqrt{\frac{3}{8\pi}}\left(\frac{i}{r} - \frac{x+iy}{r^2}\frac{y}{r}\right)\\ &= i\hbar\sqrt{\frac{3}{8\pi}}\frac{1}{r^3}\left(ir^2 - xy - iy^2\right),\end{aligned}$$

$$
\begin{aligned}
\widehat{L}_z\vec{Y}_{1,1} &:= (x\widehat{p}_y - y\widehat{p}_x)\,Y_{1,1} \\
&= i\hbar\sqrt{\frac{3}{8\pi}}\frac{1}{r^3}\Big\{\left(ixr^2 - x^2y - ixy^2\right) \\
&= -\left(yr^2 - yx^2 - ixy^2\right)\Big\} \\
&= i\hbar\sqrt{\frac{3}{8\pi}}\frac{1}{r^3}\left\{ixr^2 - yr^2\right\} \\
&= -\hbar\sqrt{\frac{3}{8\pi}}\frac{1}{r}(x+iy) = \hbar\,Y_{1,1}.
\end{aligned}
$$

SQ36(1)(b)

(1) Consider a transformation from the original coordinate system (x, y, z) to a new coordinate system (x', y', z'):

$$x \to z', \quad y \to x', \quad z \to y'.$$

We can see that the new z'-axis coincides with the original x-axis, and that

$$r = \sqrt{x^2 + y^2 + z^2} = \sqrt{x'^2 + y'^2 + z'^2} = r'.$$

Physically we expect the angular momentum about the x-axis to be the same as the angular momentum about the z'-axis. We can confirm this mathematically by showing that

$$\widehat{L}_x = y\widehat{p}_z - z\widehat{p}_y = -i\hbar\left(y\frac{\partial}{\partial z} - z\frac{\partial}{\partial y}\right)$$

is equal to

$$\widehat{L}_{z'} = x'\widehat{p}_{y'} - y'\widehat{p}_{x'} = -i\hbar\left(x'\frac{\partial}{\partial y'} - y'\frac{\partial}{\partial x'}\right).$$

This is obvious since $x' = y, y' = z$. It follows that the eigenfunctions and eigenvalues of $\widehat{L}_x$ are the same in terms of the dashed coordinates as that of $\widehat{L}_{z'}$.

(2) We know the eigenfunctions $Y'_{\ell,m_\ell}(x', y', z')$ and eigenvalues of $\widehat{L}_{z'}$ in the transformed coordinates (x', y', z') as they are of the same form as those of the eigenfunctions and eigenvalues of $\widehat{L}_z$ in the original coordinates, i.e., we have

$$Y'_{1,-1}(x', y', z') = \sqrt{\frac{3}{8\pi}}\frac{x' - iy'}{r'}, \quad Y'_{1,0}(x', y', z') = \sqrt{\frac{3}{4\pi}}\frac{z'}{r'},$$

$$Y'_{1,1}(x', y', z') = -\sqrt{\frac{3}{8\pi}}\frac{x' + iy'}{r'}.$$

Since $\widehat{L}_{z'} = \widehat{L}_x$ these are also eigenfunctions of $\widehat{L}_x$. Written in terms of the original coordinates these functions become

$$Y'_{1,-1} = \sqrt{\frac{3}{8\pi}}\frac{y-iz}{r}, \quad Y'_{1,0} = \sqrt{\frac{3}{4\pi}}\frac{x}{r},$$
$$Y'_{1,1} = -\sqrt{\frac{3}{8\pi}}\frac{y+iz}{r}.$$

We can conclude that the required eigenfunctions $\widehat{L}_x$, denoted more conveniently by $X_{1,1}$, $X_{1,0}$, $X_{1,-1}$ are

$$X_{1,-1} = \sqrt{\frac{3}{8\pi}}\frac{y-iz}{r},$$
$$X_{1,0} = \sqrt{\frac{3}{4\pi}}\frac{x}{r},$$
$$X_{1,1} = -\sqrt{\frac{3}{8\pi}}\frac{y+iz}{r}.$$

Q36(2) Suppose $\widehat{L}_c^2$ and $\widehat{L}_{cz}$ are measured giving the eigenvalues $2\hbar^2$ and $-\hbar$, respectively. A measurement of $\widehat{L}_{cx}$ is then made. What are the possible results of the measurement of $\widehat{L}_{cx}$? Find the probability of each of these possible results.

SQ36(2) The first measurement serves as a state preparation process, assuming the projection postulate. From the measured values of $\widehat{L}^2$ and $\widehat{L}_z$ we deduce that $\ell = 1, m_\ell = -1$. The projection postulate tells us that the state after the first measurement is described by the eigenfunction

$$Y_{1,-1} = \sqrt{\frac{3}{8\pi}}\frac{x-iy}{r}.$$

It also follows that the possible values of $\widehat{L}_x$ must be $-\hbar$, 0, $\hbar$, since the orbital angular momentum quantum number ℓ is 1. According to Postulate 28.1(PDDO) the probability of obtaining any of these results are:

$$\begin{aligned}\wp_{-1} &= \langle \vec{Y}_{1,-1} \mid \widehat{P}_{\vec{X}_{1,-1}} \vec{Y}_{1,-1}\rangle \\ &= \langle \vec{Y}_{1,-1} \mid \left(|\vec{X}_{1,-1}\rangle\langle\vec{X}_{1,-1}|\right) \vec{Y}_{1,-1}\rangle \\ &= \langle \vec{Y}_{1,-1} \mid \vec{X}_{1,-1}\rangle \langle \vec{X}_{1,-1} \mid \vec{Y}_{1,-1}\rangle \\ &= \left|\langle \vec{X}_{1,-1} \mid \vec{Y}_{1,-1}\rangle\right|^2.\end{aligned}$$

$$\begin{aligned}
\wp_0 &= \langle \vec{Y}_{1,-1} \mid \widehat{P}_{\vec{X}_{1,0}} \vec{Y}_{1,-1} \rangle \\
&= \langle \vec{Y}_{1,-1} \mid \left(|\vec{X}_{1,0}\rangle\langle\vec{X}_{1,0}| \right) \vec{Y}_{1,-1} \rangle \\
&= \langle \vec{Y}_{1,-1} \mid \vec{X}_{1,0} \rangle \, \langle \vec{X}_{1,0} \mid \vec{Y}_{1,-1} \rangle \\
&= \left| \langle \vec{X}_{1,0} \mid \vec{Y}_{1,-1} \rangle \right|^2 . \\
\wp_1 &= \langle \vec{Y}_{1,-1} \mid \widehat{P}_{\vec{X}_{1,1}} \vec{Y}_{1,-1} \rangle \\
&= \langle \vec{Y}_{1,-1} \mid \left(|\vec{X}_{1,1}\rangle\langle\vec{X}_{1,1}| \right) \vec{Y}_{1,-1} \rangle \\
&= \langle \vec{Y}_{1,-1} \mid \vec{X}_{1,1} \rangle \, \langle \vec{X}_{1,1} \mid \vec{Y}_{1,-1} \rangle \\
&= \left| \langle \vec{X}_{1,1} \mid \vec{Y}_{1,-1} \rangle \right|^2 .
\end{aligned}$$

It is tedious to evaluate all these scalar products. A simpler way to proceed is to recognise that if we expand the initial state $\vec{Y}_{1,-1}$ in terms of the eigenvectors of $\widehat{L}_x$, i.e.,

$$\vec{Y}_{1,-1} = c_{-1}\vec{X}_{1,-1} + c_0\vec{X}_{1,0} + c_1\vec{X}_{1,1},$$

then according to Eq. (28.4) the desired probabilities are equal to the absolute value squares of the coefficients of expansion, i.e.,

$$\wp_{-1} = |c_{-1}|^2, \quad \wp_0 = |c_0|^2, \quad \wp_1 = |c_1|^2.$$

Explicitly we have

$$\begin{aligned}
&\sqrt{\frac{3}{8\pi}}\,\frac{x-iy}{r} \\
&= c_{-1}\sqrt{\frac{3}{8\pi}}\,\frac{y-iz}{r} + c_0\sqrt{\frac{3}{4\pi}}\,\frac{x}{r} + c_1\left(-\sqrt{\frac{3}{8\pi}}\,\frac{y+iz}{r}\right). \\
\Rightarrow\;\; & x - iy = c_{-1}(y - iz) + c_0\sqrt{2}\,x - c_1(y + iz) \\
& \qquad = \sqrt{2}\,c_0 x + (c_{-1} - c_1)y - i(c_{-1} + c_1)z.
\end{aligned}$$

Now, equate the coefficients of x, y, and z on both sides in turn we can obtain the values of c_1, c_0, and c_{-1}:

$$1 = \sqrt{2}c_0, \quad -i = c_{-1} - c_1, \quad 0 = c_{-1} + c_1$$

It follows that

$$c_0 = 1/\sqrt{2}, \;\; c_{-1} = -i/2, \;\; c_1 = i/2,$$

and

$$\wp_{-1} = |c_{-1}|^2 = \frac{1}{4}, \quad \wp_0 = |c_0|^2 = \frac{1}{2}, \quad \wp_1 = |c_1|^2 = \frac{1}{4}.$$

The total probability is 1 as it must be.

Q36(3) The Hamiltonian of a classical particle of mass m constrained to move freely on the surface of a sphere of radius a is

$$H = \frac{1}{2I} L^2,$$

where $I = ma^2$ is the moment of inertia of the particle and L^2 is the total orbital angular momentum square of the particle, both with respect to the origin. Quantise the system and find the energy eigenvalues and the corresponding eigenfunctions of the quantised system. What are the degeneracy of the energy eigenvalues?[3]

SQ36(3) The state space of the system (a rigid rotator) is the Hilbert space $\vec{L}^2(\mathcal{S}_u)$ introduced in §16.1.2.8 and in E16.2.2(3), i.e., it is defined by the set $L^2(\mathcal{S}_u)$ of square-integrable functions defined on the unit sphere $\mathcal{S}_u$ centered at the coordinate origin. This set is spanned by the spherical harmonics $Y_{\ell,m_\ell}(\theta, \varphi)$ first introduced in Eq. (16.63).[4] Following the discussion in §36.1.1 the square of the orbital angular momentum of the rigid rotator is quantised as the operator $\widehat{L^2}(\mathcal{S}_u)$ in Eq. (19.50).[5] It follows that the Hamiltonian is quantised as the operator $\widehat{H} = \widehat{L^2}(\mathcal{S}_u)/2I$. The eigenfunctions are the spherical harmonics $Y_{\ell,m_\ell}(\theta, \varphi)$ corresponding to eigenvalue

$$E_{\ell,m_\ell} = \ell(\ell+1)\hbar^2/2I, \quad \ell = 0, 1, 2, \cdots.$$

Apart from the eigenvalue $E_{0,0}$ all the eigenvalues are degenerate with degeneracy equal to the number of different m_ℓ values for a given ℓ, i.e., $2\ell + 1$ since m_ℓ ranges from $-\ell$ to ℓ.

---

[3]Zettili pp. 296–297. Such a system is known as a *rigid rotator* which can be used to model a diatomic molecule. For the state space of a quantum rigid rotator, see the comment on a footnote in §27.8.

[4]A general and explicit expression of $Y_{\ell,m_\ell}(\theta, \varphi)$ is given by Eq. (36.42).

[5]*See* Eq. (36.8) and the discussion there.

Q36(4) Verify the commutation relation in Eq. (36.51).

SQ36(4) Using the linearity and product rules for commutators

$$[\widehat{A}, \widehat{B}+\widehat{C}\,] = [\widehat{A}, \widehat{B}\,] + [\widehat{A}, \widehat{C}\,],$$

$$[\widehat{A}, \widehat{B}\widehat{C}\,] = [\widehat{A}, \widehat{B}\,]\widehat{C} + \widehat{B}\,[\widehat{A}, \widehat{C}\,],$$

and the commutation relations of the annihilation and creation operators, e.g.,

$$\hat{a}_1^\dagger\hat{a}_2\hat{a}_2^\dagger\hat{a}_1 = \hat{a}_1^\dagger\hat{a}_1\hat{a}_2\hat{a}_2^\dagger = \hat{a}_1^\dagger\hat{a}_1\left(\hat{a}_2^\dagger\hat{a}_2+1\right) = \widehat{N}_1(\widehat{N}_2+1),$$

we can calculate the commutation relation $[\widehat{J}_x, \widehat{J}_y\,]$ as follows:

$$\begin{aligned}
[\widehat{J}_x, \widehat{J}_y\,] &= \frac{\hbar^2}{4i}\,[\,(\hat{a}_1^\dagger\hat{a}_2+\hat{a}_2^\dagger\hat{a}_1), (\hat{a}_1^\dagger\hat{a}_2-\hat{a}_2^\dagger\hat{a}_1)\,] \\
&= \frac{\hbar^2}{4i}\Big\{[\,(\hat{a}_1^\dagger\hat{a}_2, (\hat{a}_1^\dagger\hat{a}_2-\hat{a}_2^\dagger\hat{a}_1)\,] \\
&\quad + [\,\hat{a}_2^\dagger\hat{a}_1, (\hat{a}_1^\dagger\hat{a}_2-\hat{a}_2^\dagger\hat{a}_1)\,]\Big\} \\
&= \frac{\hbar^2}{4i}\Big\{-[\,\hat{a}_1^\dagger\hat{a}_2, \hat{a}_2^\dagger\hat{a}_1\,] + [\,\hat{a}_2^\dagger\hat{a}_1, \hat{a}_1^\dagger\hat{a}_2\,]\Big\} \\
&= \frac{\hbar^2}{4i}\Big\{\Big(-\hat{a}_1^\dagger\,[\,\hat{a}_2, \hat{a}_2^\dagger\hat{a}_1\,] - [\,\hat{a}_1^\dagger, \hat{a}_2^\dagger\hat{a}_1\,]\,\hat{a}_2\Big) \\
&\quad + \Big(\hat{a}_2^\dagger\,[\,\hat{a}_1, \hat{a}_1^\dagger\hat{a}_2\,] + [\,\hat{a}_2^\dagger, \hat{a}_1^\dagger\hat{a}_2\,]\,\hat{a}_1\Big)\Big\} \\
&= \frac{\hbar^2}{4i}\Big\{\Big(-\hat{a}_1^\dagger\hat{a}_1+\hat{a}_2^\dagger\hat{a}_2\Big) + \Big(\hat{a}_2^\dagger\hat{a}_2-\hat{a}_1^\dagger\hat{a}_1\Big)\Big\} \\
&= \frac{\hbar^2}{2i}\Big(\hat{a}_2^\dagger\hat{a}_2-\hat{a}_1^\dagger\hat{a}_1\Big) = i\hbar\widehat{J}_z.
\end{aligned}$$

Q36(5) Verify Eqs. (36.54), (36.55) and (36.60).

SQ36(5) To verify Eq. (36.54) we only need to observe that $\widehat{N}_1 = \hat{a}_1^\dagger\hat{a}_1$ and $\widehat{N}_2 = \hat{a}_2^\dagger\hat{a}_2$. To verify Eq. (36.55) we need to express $\widehat{J}_x^{\,2}, \widehat{J}_y^{\,2}$ and $\widehat{J}_z^{\,2}$ in terms of the annihilation and creation operators, i.e.,

$$\begin{aligned}
\widehat{J}_x^{\,2} &= \frac{\hbar^2}{4}\left((\hat{a}_1^\dagger\hat{a}_2)^2+(\hat{a}_2^\dagger\hat{a}_1)^2+\hat{a}_1^\dagger\hat{a}_2\hat{a}_2^\dagger\hat{a}_1+\hat{a}_2^\dagger\hat{a}_1\hat{a}_1^\dagger\hat{a}_2\right) \\
&= \frac{\hbar^2}{4}\left((\hat{a}_1^\dagger\hat{a}_2)^2+(\hat{a}_2^\dagger\hat{a}_1)^2+\widehat{N}_1(\widehat{N}_2+1)+\widehat{N}_2(\widehat{N}_1+1)\right),
\end{aligned}$$

$$\widehat{J}_y^{\,2} = -\frac{\hbar^2}{4}\Big((\widehat{a}_1^\dagger\widehat{a}_2)^2 + (\widehat{a}_2^\dagger\widehat{a}_1)^2 - \widehat{N}_1(\widehat{N}_2+1) - \widehat{N}_2(\widehat{N}_1+1)\Big),$$

$$\widehat{J}_z^{\,2} = \frac{\hbar^2}{4}\Big(\widehat{N}_1^2 + \widehat{N}_2^2 - \widehat{N}_1\widehat{N}_2 - \widehat{N}_2\widehat{N}_1\Big),$$

$$\Rightarrow\ \widehat{J}^{\,2} = \frac{\hbar^2}{4}\Big(\widehat{N}_1^2 + \widehat{N}_2^2 + \widehat{N}_1\widehat{N}_2 + \widehat{N}_2\widehat{N}_1 + 2\widehat{N}_1 + 2\widehat{N}_2\Big)$$

$$= \frac{\hbar^2}{4}\Big(\big(\widehat{N}_1 + \widehat{N}_2\big)^2 + 2\big(\widehat{N}_1 + \widehat{N}_2\big)\Big)$$

$$= \frac{\widehat{N}}{2}\left(\frac{\widehat{N}}{2} + 1\right)\hbar^2.$$

To verify Eq. (36.60) we observe that

$$\widehat{J}_+ = \widehat{J}_x + i\widehat{J}_y = \frac{\hbar}{2}\Big((\widehat{a}_1^\dagger\widehat{a}_2 + \widehat{a}_2^\dagger\widehat{a}_1) + (\widehat{a}_1^\dagger\widehat{a}_2 - \widehat{a}_2^\dagger\widehat{a}_1)\Big)$$

$$= \hbar\,\widehat{a}_1^\dagger\widehat{a}_2.$$

$$\widehat{J}_- = \widehat{J}_x - i\widehat{J}_y = \frac{\hbar}{2}\Big((\widehat{a}_1^\dagger\widehat{a}_2 + \widehat{a}_2^\dagger\widehat{a}_1) - (\widehat{a}_1^\dagger\widehat{a}_2 - \widehat{a}_2^\dagger\widehat{a}_1)\Big)$$

$$= \hbar\,\widehat{a}_2^\dagger\widehat{a}_1.$$

Q36(6) Verify properties P36.2.2(1), P36.2.2(2) and P36.2.2(3). including Eqs. (36.73) and (36.74).

SQ36(6) The starting point is Eq. (36.62).

(1) To prove P36.2.2(1) we first note that n_1 and n_2 takes 0 and positive integers values. Consequently $m_j = (n_1 - n_2)/2$ would take integer and half integer values. It is positive when $n_1 > n_2$ and negative when $n_1 < n_2$. It takes the value 0 when $n_1 = n_2$. Since $j = (n_1 + n_2)/2$ it can only take the value 0, positive half integer and integer values.

(2) To prove P36.2.2(2) we have

$$m^2 = \frac{1}{4}\left(n_1^2 - 2n_1n_2 + n_2^2\right),\quad j^2 = \frac{1}{4}\left(n_1^2 + 2n_1n_2 + n_2^2\right)$$

$$\Rightarrow\ m^2 \le j^2 \quad \Rightarrow\ -j \le m \le j,$$

since n_1 and n_2 are both non-negative. If either n_1 or n_2 is zero then $J^2 = m^2$. Since $m = (n_1 - n_2)/2$ is negative when $n_1 = 0$ we have $m = -j$ and when $n_1 = 0$ and $m = j$ if $n_2 = 0$.

(3) To prove P36.2.2(3) we use the expressions

$$|j, m\rangle = \vec{\varphi}_{n_1,n_2}, \quad \widehat{J}_+ = \hbar \widehat{a}_1^\dagger \widehat{a}_2, \quad \widehat{J}_- = \hbar \widehat{a}_2^\dagger \widehat{a}_1.$$

We get

$$\widehat{J}_+ |j, m\rangle = \hbar \widehat{a}_1^\dagger \widehat{a}_2 \, \vec{\varphi}_{n_1,n_2} = \hbar \sqrt{(n_1 + 1)\, n_2} \, \vec{\varphi}_{n_1+1,n_2-1}. \qquad (*)$$

Next we desire to express $\vec{\varphi}_{n_1+1,n_2-1}$ in the notation $|j', m'\rangle$. This can be done by using Eq. (36.62) and rewriting $\vec{\varphi}_{n_1+1,n_2-1}$ as $\vec{\varphi}_{n_1',n_2'}$, where $n_1' = n_1 + 1$ and $n_2' = n_2 - 1$. Then we have:

$$\begin{aligned} j' &= \frac{1}{2}\,(n_1' + n_2') = \frac{1}{2}\big((n_1 + 1) + (n_2 - 1)\big) = j, \\ m' &= \frac{1}{2}\,(n_1' - n_2') = \frac{1}{2}\big((n_1 + 1) - (n_2 - 1)\big) = m + 1, \\ \Rightarrow \quad & \vec{\varphi}_{n_1+1,n_2-1} = |j, m + 1\rangle. \end{aligned}$$

Equation $(*)$ becomes

$$\widehat{J}_+ |j, m\rangle = \hbar \sqrt{(n_1 + 1)\, n_2} \; |j, m + 1\rangle. \qquad (**)$$

Since $j = (n_1 + n_2)/2$ and $m = (n_1 - n_2)/2$ we get $n_2 = j - m$ and $n_1 = j + m$, i.e., we have

$$\sqrt{n_2\,(n_1 + 1)} = \sqrt{(j - m)(j + m + 1)}.$$

Equation $(**)$ can be rewritten in the form of Eq. (36.73), i.e.,

$$\widehat{J}_+ |j, m\rangle = \hbar \sqrt{(j - m)(j + m + 1)} \; |j, m + 1\rangle.$$

Equation (36.74) is proved similarly, i.e., we have

$$\widehat{J}_- |j, m\rangle = \hbar \widehat{a}_2^\dagger \widehat{a}_1 \, \vec{\varphi}_{n_1,n_2} = \hbar \sqrt{(n_2 + 1)\, n_1} \, \vec{\varphi}_{n_1-1,n_2+1}.$$

Writing $\vec{\varphi}_{n_1-1,n_2+1}$ in the notation $|j'', m''\rangle$ and using Eq. (36.62) we get

$$\begin{aligned} j'' &= \frac{1}{2}\,(n_1'' + n_2'') = \frac{1}{2}\big((n_1 - 1) + (n_2 + 1)\big) = j, \\ m'' &= \frac{1}{2}\,(n_1'' - n_2'') = \frac{1}{2}\big((n_1 - 1) - (n_2 + 1)\big) = m - 1, \\ & \sqrt{n_1\,(n_2 + 1)} = \sqrt{(j + m)(j - m + 1)}, \end{aligned}$$

which immediately leads to Eq. (36.74).

Q36(7) How is the vector $\vec{\Phi}$ in Eq. (36.108) normalised?

SQ36(7) First we note that the functions $\phi_+(\vec{x})$ and $\phi_-(\vec{x})$ are normalised separately in accordance of Eq. (36.89), and that the spin vectors $\vec{\alpha}$ and $\vec{\beta}$ are orthonormal by definition. The vector $\vec{\Phi}$ is then normalised by requiring

$$\langle \vec{\Phi} \mid \vec{\Phi} \rangle = |c_+|^2 \int_{-\infty}^{\infty} |\phi_+(\vec{x})|^2 \, dxdydz$$
$$|c_-|^2 \int_{-\infty}^{\infty} |\phi_-(\vec{x})|^2 \, dxdydz$$
$$= |c_+|^2 + |c_-|^2 = 1.$$

Q36(8) Prove Eqs. (36.124) and (36.125) directly using the defining properties of spin operators given in P36.3(1) to P36.3(4) without using Eqs. (36.82), (36.83) and (36.84).

SQ36(8) The answers are already given in item 8 of §36.3.5. First we can construct the operators $\widehat{S}_\pm = \widehat{S}_x \pm i\widehat{S}_y$. These are operators on $\vec{\mathbb{V}}^2$. Then Eqs. (36.13) and (36.17) apply to spin operators since these equations are derived using only the commutation relations of the operators concerned. For examples we have

$$\widehat{S}_+\widehat{S}_- = \widehat{S}^2 - \widehat{S}_z^2 + \hbar\widehat{S}_z \quad \text{and} \quad \widehat{S}_-\widehat{S}_+ = \widehat{S}^2 - \widehat{S}_z^2 - \hbar\widehat{S}_z$$
$$\widehat{S}_z\widehat{S}_+ = \widehat{S}_+\widehat{S}_z + \hbar\widehat{S}_+ \quad \text{and} \quad \widehat{S}_z\widehat{S}_- = \widehat{S}_-\widehat{S}_z - \hbar\widehat{S}_-.$$

Equation (36.124) is proved as follows:

(1) Applying $\widehat{S}_z\widehat{S}_+ = \widehat{S}_+\widehat{S}_z + \hbar\widehat{S}_+$ to $\vec{\beta}_z$ we get

$$\widehat{S}_z\widehat{S}_+\vec{\beta}_z = \big(\widehat{S}_+\widehat{S}_z + \hbar\widehat{S}_+\big)\vec{\beta}_z = \left(-\frac{1}{2}\hbar + \hbar\right)\widehat{S}_+\vec{\beta}_z$$
$$= \frac{1}{2}\hbar\,\widehat{S}_+\vec{\beta}_z.$$

This means that $\widehat{S}_+\vec{\beta}_z$ is an eigenvector of $\widehat{S}_z$ corresponding to eigenvalue $\hbar/2$ which is nondegenerate. It follows that we must have $\widehat{S}_+\vec{\beta}_z = c_+\vec{\alpha}_z$. We then have $\langle \widehat{S}_+\vec{\beta}_z \mid \widehat{S}_+\vec{\beta}_z \rangle = |c_+|^2$. The constant c_+ can be calculated as follows:

$$|c_+|^2 = \langle \widehat{S}_+\vec{\beta}_z \mid \widehat{S}_+\vec{\beta}_z \rangle = \langle \widehat{S}_-\widehat{S}_+\vec{\beta}_z \mid \vec{\beta}_z \rangle$$
$$= \langle \big(\widehat{S}^2 - \widehat{S}_z^2 - \hbar\widehat{S}_z\big)\vec{\beta}_z \mid \vec{\beta}_z \rangle$$
$$= \langle \Big(\frac{3}{4} - \frac{1}{4} + \frac{1}{2}\Big)\hbar^2\vec{\beta}_z \mid \vec{\beta}_z \rangle = \hbar^2.$$

Choosing $c_+ = \hbar$ gives the answer.[6]

(2) Applying $\widehat{S}_z\widehat{S}_- = \widehat{S}_-\widehat{S}_z - \hbar\widehat{S}_-$ to $\vec{\alpha}_z$ we get

$$\widehat{S}_z\widehat{S}_-\vec{\alpha}_z = (\widehat{S}_-\widehat{S}_z - \hbar\widehat{S}_-)\vec{\alpha}_z = \left(\frac{1}{2}\hbar - \hbar\right)\widehat{S}_-\vec{\alpha}_z = -\frac{1}{2}\hbar\,\widehat{S}_-\vec{\alpha}_z.$$

This means that $\widehat{S}_-\vec{\alpha}_z$ is an eigenvector of $\widehat{S}_z$ corresponding to eigenvalue $-\hbar/2$ which is nondegenerate. It follows that we must have $\widehat{S}_-\vec{\alpha}_z = c_-\vec{\beta}_z$. The constant c_- can be calculated as follows:

$$\begin{aligned}|c_-|^2 &= \langle \widehat{S}_-\vec{\alpha}_z \mid \widehat{S}_-\vec{\alpha}_z\rangle = \langle \widehat{S}_+\widehat{S}_-\vec{\alpha}_z \mid \vec{\alpha}_z\rangle \\ &= \langle(\widehat{S}^2 - \widehat{S}_z^2 + \hbar\widehat{S}_z))\vec{\alpha}_z \mid \vec{\alpha}_z\,\rangle \\ &= \langle(\frac{3}{4} - \frac{1}{4} + \frac{1}{2})\hbar^2\vec{\alpha}_z \mid \vec{\alpha}_z\,\rangle = \hbar^2.\end{aligned}$$

Choosing $c_- = \hbar$ gives the answer.

Equation (36.125) is proved as follows:

(1) Applying $\widehat{S}_z\widehat{S}_+ = \widehat{S}_+\widehat{S}_z + \hbar\widehat{S}_+$ to $\vec{\alpha}_z$ we get

$$\begin{aligned}\widehat{S}_z\widehat{S}_+\vec{\alpha}_z &= (\widehat{S}_+\widehat{S}_z + \hbar\widehat{S}_+)\vec{\alpha}_z = \left(\frac{1}{2}\hbar + \hbar\right)\widehat{S}_+\vec{\alpha}_z \\ &= \frac{3}{2}\hbar\,\widehat{S}_+\vec{\alpha}_z.\end{aligned}$$

This means that $\widehat{S}_+\vec{\alpha}_z$ is an eigenvector of $\widehat{S}_z$ corresponding to the eigenvalue $3\hbar/2$. This is a contradiction since by definition $\widehat{S}_z$ does not have $3\hbar/2$ as an eigenvalue. To avoid the contradiction we must have $\widehat{S}_+\vec{\alpha}_z = \vec{0}$. Another way of looking at this is that if $\widehat{S}_+\vec{\alpha}_z \neq \vec{0}$ then it would be linear independent of $\vec{\alpha}_z$ and $\vec{\beta}_z$. This would make the state space three-dimensional, contradicting defining property P36.3(2). So, we can conclude that $\widehat{S}_+\vec{\alpha}_z = \vec{0}$.

(2) Applying the result $\widehat{S}_z\widehat{S}_- = \widehat{S}_-\widehat{S}_z - \hbar\widehat{S}_-$ to $\vec{\beta}_z$ we get

$$\begin{aligned}\widehat{S}_z\widehat{S}_-\vec{\beta}_z &= (\widehat{S}_-\widehat{S}_z - \hbar\widehat{S}_+)\vec{\beta}_z = \left(-\frac{1}{2}\hbar - \hbar\right)\widehat{S}_-\vec{\beta}_z \\ &= -\frac{3}{2}\hbar\,\widehat{S}_-\vec{\beta}_z.\end{aligned}$$

[6] *See* Merzbacher p. 240 for comments on this choice.

This means that $\widehat{S}_{-}\vec{\beta}_z$ is an eigenvector of $\widehat{S}_z$ corresponding to the eigenvalue $-3\hbar/2$. This is a contradiction since by definition $\widehat{S}_z$ does not have $-3\hbar/2$ as an eigenvalue. To avoid the contradiction we must have $\widehat{S}_{-}\vec{\beta}_z = \vec{0}$.

Q36(9) Using the matrix representation of $\widehat{S}_{(1)x}$, $\widehat{S}_{(1)y}$, $\widehat{S}_{(1)z}$ for a spin-1 particle in Eqs. (14.48), (14.49) and (14.50) show that the matrices for

$$\widehat{S}_{(1)+} = \widehat{S}_{(1)x} + i\widehat{S}_{(1)y}, \quad \widehat{S}_{(1)-} = \widehat{S}_{(1)x} - i\widehat{S}_{(1)y}, \quad \widehat{S}^2_{(1)},$$

are

$$\boldsymbol{M}_{\hat{S}_{(1)+}} = \sqrt{2}\hbar \begin{pmatrix} 0 & 1 & 0 \\ 0 & 0 & 1 \\ 0 & 0 & 0 \end{pmatrix},$$

$$\boldsymbol{M}_{\hat{S}_{(1)-}} = \sqrt{2}\hbar \begin{pmatrix} 0 & 0 & 0 \\ 1 & 0 & 0 \\ 0 & 1 & 0 \end{pmatrix},$$

$$\boldsymbol{M}_{\hat{S}^2_{(1)}} = 2\hbar^2 \begin{pmatrix} 1 & 0 & 0 \\ 0 & 1 & 0 \\ 0 & 0 & 1 \end{pmatrix}.$$

SQ36(9) Using the expressions for $\boldsymbol{M}_{\hat{S}_{(1)x}}$ and $\boldsymbol{M}_{\hat{S}_{(1)y}}$ in Eqs. (14.48) and (14.49) we get

$$\boldsymbol{M}_{\hat{S}_{(1)+}} = \boldsymbol{M}_{\hat{S}_{(1)x}} + i\boldsymbol{M}_{\hat{S}_{(1)y}} = \sqrt{2}\hbar \begin{pmatrix} 0 & 1 & 0 \\ 0 & 0 & 1 \\ 0 & 0 & 0 \end{pmatrix},$$

$$\boldsymbol{M}_{\hat{S}_{(1)-}} = \boldsymbol{M}_{\hat{S}_{(1)x}} - i\boldsymbol{M}_{\hat{S}_{(1)y}} = \sqrt{2}\hbar \begin{pmatrix} 0 & 0 & 0 \\ 1 & 0 & 0 \\ 0 & 1 & 0 \end{pmatrix}.$$

For $\boldsymbol{M}_{\hat{S}^2_{(1)}}$ we have to evaluate $\boldsymbol{M}_{\hat{S}^2_{(1)x}}$, $\boldsymbol{M}_{\hat{S}^2_{(1)y}}$ and $\boldsymbol{M}_{\hat{S}^2_{(1)z}}$ first. We have

$$\boldsymbol{M}^2_{\hat{S}_{(1)x}} = \frac{\hbar^2}{2} \begin{pmatrix} 0 & 1 & 0 \\ 1 & 0 & 1 \\ 0 & 1 & 0 \end{pmatrix}^2 = \frac{\hbar^2}{2} \begin{pmatrix} 1 & 0 & 1 \\ 0 & 2 & 0 \\ 1 & 0 & 1 \end{pmatrix},$$

$$\boldsymbol{M}^2_{\hat{S}_{(1)y}} = \frac{\hbar^2}{2} \begin{pmatrix} 0 & -i & 0 \\ i & 0 & -i \\ 0 & i & 0 \end{pmatrix}^2 = \frac{\hbar^2}{2} \begin{pmatrix} 1 & 0 & -1 \\ 0 & 2 & 0 \\ -1 & 0 & 1 \end{pmatrix},$$

$$\boldsymbol{M}^2_{\hat{S}_{(1)z}} = \hbar^2 \begin{pmatrix} 1 & 0 & 0 \\ 0 & 0 & 0 \\ 0 & 0 & -1 \end{pmatrix}^2 = \hbar^2 \begin{pmatrix} 1 & 0 & 0 \\ 0 & 0 & 0 \\ 0 & 0 & 1 \end{pmatrix}$$

$$= \frac{\hbar^2}{2} \begin{pmatrix} 2 & 0 & 0 \\ 0 & 0 & 0 \\ 0 & 0 & 2 \end{pmatrix}.$$

It follows that

$$\boldsymbol{M}_{\hat{S}^2_{(1)}} = \boldsymbol{M}^2_{\hat{S}_{(1)x}} + \boldsymbol{M}^2_{\hat{S}_{(1)y}} + \boldsymbol{M}^2_{\hat{S}_{(1)z}} = 2\hbar^2 \begin{pmatrix} 1 & 0 & 0 \\ 0 & 1 & 0 \\ 0 & 0 & 1 \end{pmatrix}.$$

Q36(10) Using Eqs. (36.126) and (36.127) verify that $\vec{\alpha}_x$, $\vec{\beta}_x$ defined by Eq. (36.137) are eigenvectors of $\widehat{S}_x$ and $\vec{\alpha}_y$ and $\vec{\beta}_y$ defined by Eq. (36.138) are eigenvectors of $\widehat{S}_y$.

SQ36(10) Using the results of $\widehat{S}_x$ and $\widehat{S}_y$ acting on $\vec{\alpha}_z$ and $\vec{\beta}_z$ in Eqs. (36.126) and (36.127) we get

$$\widehat{S}_x\,\vec{\alpha}_x = \frac{1}{\sqrt{2}}\left(\widehat{S}_x\,\vec{\alpha}_z + \widehat{S}_x\,\vec{\beta}_z\right) = \frac{1}{2}\hbar\,\frac{1}{\sqrt{2}}\,(\vec{\beta}_z + \vec{\alpha}_z)$$

$$= \frac{1}{2}\,\hbar\,\vec{\alpha}_x.$$

$$\widehat{S}_x\,\vec{\beta}_x = \frac{1}{\sqrt{2}}\left(\widehat{S}_x\,\vec{\alpha}_z - \widehat{S}_x\,\vec{\beta}_z\right) = \frac{1}{2}\hbar\,\frac{1}{\sqrt{2}}\,(\vec{\beta}_z - \vec{\alpha}_z)$$

$$= -\frac{1}{2}\,\hbar\,\vec{\beta}_x.$$

$$\widehat{S}_y\,\vec{\alpha}_y = \frac{1}{\sqrt{2}}\left(\widehat{S}_y\,\vec{\alpha}_z + i\widehat{S}_y\,\vec{\beta}_z\right) = \frac{1}{2}\hbar\,\frac{1}{\sqrt{2}}\,(i\vec{\beta}_z + \vec{\alpha}_z)$$

$$= \frac{1}{2}\,\hbar\,\vec{\alpha}_y.$$

$$\widehat{S}_y\,\vec{\beta}_y = \frac{1}{\sqrt{2}}\left(\widehat{S}_y\,\vec{\alpha}_z - i\widehat{S}_y\,\vec{\beta}_z\right) = \frac{1}{2}\hbar\,\frac{1}{\sqrt{2}}\left(i\vec{\beta}_z - \vec{\alpha}_z\right)$$

$$= -\frac{1}{2}\,\hbar\,\vec{\beta}_y.$$

These results show that $\vec{\alpha}_x$ and $\vec{\beta}_x$ are eigenvectors of $\widehat{S}_x$, and $\vec{\alpha}_y$ and $\vec{\beta}_y$ are eigenvectors of $\widehat{S}_y$. The corresponding eigenvalues are $\pm\hbar/2$.

Q36(11) The z-component spin is measured giving a value $-\frac{1}{2}\hbar$. What are the possible outcomes of a measurement of the x-component spin? Find the probabilities of these possible measured outcomes.

SQ36(11) This question is similar to Q36(2). The first measurement of the z-component spin serves as a state preparation process, i.e., the state after is given by the state vector $\vec{\beta}_z$. Then the question turns to the outcomes of a measurement of the x-component spin given state vector $\vec{\beta}_z$. There are always two possible outcomes for a spin measurement, i.e., possible outcomes of a measurement of the x-component spin are $\pm\hbar/2$. To find the probabilities let $\widehat{P}_{\vec{\alpha}_x} = |\vec{\alpha}_x\rangle\langle\vec{\alpha}_x|$ be the projector generated by $\vec{\alpha}_x$. Then Postulate 28.1(PDDO) tells us that the probability of a measurement of $\widehat{S}_x$ given state $\vec{\beta}_z$ yield the value $+\hbar/2$ is

$$\wp_{x,+} = \langle\vec{\beta}_z \mid \widehat{P}_x\vec{\beta}_z\rangle = \langle\vec{\beta}_z \mid \left(|\vec{\alpha}_x\rangle\langle\vec{\alpha}_x|\right)\vec{\beta}_z\rangle$$

$$= \langle\vec{\beta}_z \mid \vec{\alpha}_x\rangle\,\langle\vec{\alpha}_x \mid \vec{\beta}_z\rangle = |\langle\vec{\beta}_z \mid \vec{\alpha}_x\rangle\,|^2.$$

Using the expression for $\vec{\alpha}_x$ in Eq. (36.139) and the orthonormality of $\vec{\alpha}_z$ and $\vec{\beta}z$ the scalar product $\langle\vec{\beta}_z \mid \vec{\alpha}_x\rangle$ is easily calculated, i.e., we have

$$\langle\vec{\beta}_z \mid \vec{\alpha}_x\rangle = \langle\vec{\beta}_z \mid \frac{1}{\sqrt{2}}\left(\vec{\alpha}_z + \vec{\beta}_z\right)\rangle$$

$$= \frac{1}{\sqrt{2}} \quad\Rightarrow\quad \wp_{x,+} = \frac{1}{2}.$$

The probability for the value $-\hbar/2$ is obtained in the same way, i.e., we have

$$\langle\vec{\beta}_z \mid \vec{\beta}_x\rangle = \langle\vec{\beta}_z \mid \frac{1}{\sqrt{2}}\left(\vec{\alpha}_z - \vec{\beta}_z\right)\rangle$$

$$= -\frac{1}{\sqrt{2}} \quad\Rightarrow\quad \wp_{x,-} = \frac{1}{2}.$$

Another approach would be to express the initial state vector $\vec{\beta}_z$ in terms of the eigenvectors of $\widehat{S}_x$, i.e., we write $\vec{\beta}_z = c_+\vec{\alpha}_x + c_-\vec{\beta}_x$. Using the expression for $\vec{\alpha}_x$ and $\vec{\beta}_x$ in Eqs. (36.137) and (36.138) we can work out c_+ and c_- immediately, by equating the coefficients of

$\vec{\alpha}_z$ and $\vec{\beta}_z$ on both sides of the above equation, i.e., we have

$$\begin{aligned}\vec{\beta}_z &= c_+ \frac{1}{\sqrt{2}}(\vec{\alpha}_z + \vec{\beta}_z) + c_- \frac{1}{\sqrt{2}}(\vec{\alpha}_z - \vec{\beta}_z) \\ &= \frac{1}{\sqrt{2}}(c_+ + c_-)\vec{\alpha}_z + \frac{1}{\sqrt{2}}(c_+ - c_-)\vec{\beta}_z \\ \Rightarrow &\quad (c_+ + c_-) = 0, \quad \frac{1}{\sqrt{2}}(c_+ - c_-) = 1 \\ \Rightarrow &\quad c_+ = 1/\sqrt{2}, \;\; c_- = -1/\sqrt{2}.\end{aligned}$$

The required probabilities are $|c_+|^2 = 1/2$ and $|c_-|^2 = 1/2$ which agree with previous results.

Q36(12) Verify that Eqs. (36.146), (36.148) and (37.147) are satisfied by the matrices in Eqs. (36.143), (36.145) and (36.144).

SQ36(12) First we have, from Eqs. (36.139) to (36.141),

$$\boldsymbol{C}_{\vec{\alpha}_z} = \begin{pmatrix} 1 \\ 0 \end{pmatrix}, \qquad \boldsymbol{C}_{\vec{\beta}_z} = \begin{pmatrix} 0 \\ 1 \end{pmatrix};$$

$$\boldsymbol{C}_{\vec{\alpha}_x} = \frac{1}{\sqrt{2}}\begin{pmatrix} 1 \\ 1 \end{pmatrix}, \quad \boldsymbol{C}_{\vec{\beta}_x} = \frac{1}{\sqrt{2}}\begin{pmatrix} 1 \\ -1 \end{pmatrix};$$

$$\boldsymbol{C}_{\vec{\alpha}_y} = \frac{1}{\sqrt{2}}\begin{pmatrix} 1 \\ i \end{pmatrix}, \quad \boldsymbol{C}_{\vec{\beta}_y} = \frac{1}{\sqrt{2}}\begin{pmatrix} 1 \\ -i \end{pmatrix}.$$

Next we have the matrix representations of $\widehat{S}_z$, $\widehat{S}_x$ and $\widehat{S}_y$ in the basis $\{\vec{\alpha}_z, \vec{\beta}_z\}$ given by Eqs. (36.143) to (36.145), i.e.,

$$\boldsymbol{M}_{\widehat{S}_z} = \frac{1}{2}\hbar\begin{pmatrix} 1 & 0 \\ 0 & -1 \end{pmatrix},$$

$$\boldsymbol{M}_{\widehat{S}_x} = \frac{1}{2}\hbar\begin{pmatrix} 0 & 1 \\ 1 & 0 \end{pmatrix},$$

$$\boldsymbol{M}_{\widehat{S}_y} = \frac{1}{2}\hbar\begin{pmatrix} 0 & -i \\ i & 0 \end{pmatrix}.$$

It is then a matter of matrix multiplication, i.e., we have:

$$M_{\hat{S}_z} C_{\vec{\alpha}_z} = \frac{\hbar}{2}\begin{pmatrix} 1 & 0 \\ 0 & -1 \end{pmatrix}\begin{pmatrix} 1 \\ 0 \end{pmatrix} = \frac{\hbar}{2}\begin{pmatrix} 1 \\ 0 \end{pmatrix}$$

$$= \frac{\hbar}{2} C_{\vec{\alpha}_z},$$

$$M_{\hat{S}_z} C_{\vec{\beta}_z} = \frac{\hbar}{2}\begin{pmatrix} 1 & 0 \\ 0 & -1 \end{pmatrix}\begin{pmatrix} 0 \\ 1 \end{pmatrix} = \frac{\hbar}{2}\begin{pmatrix} 0 \\ -1 \end{pmatrix}$$

$$= -\frac{\hbar}{2} C_{\vec{\beta}_z},$$

$$M_{\hat{S}_x} C_{\vec{\alpha}_x} = \frac{\hbar}{2\sqrt{2}}\begin{pmatrix} 0 & 1 \\ 1 & 0 \end{pmatrix}\begin{pmatrix} 1 \\ 1 \end{pmatrix} = \frac{\hbar}{2\sqrt{2}}\begin{pmatrix} 1 \\ 1 \end{pmatrix}$$

$$= \frac{\hbar}{2} C_{\vec{\alpha}_x},$$

$$M_{\hat{S}_x} C_{\vec{\beta}_x} = \frac{\hbar}{2\sqrt{2}}\begin{pmatrix} 0 & 1 \\ 1 & 0 \end{pmatrix}\begin{pmatrix} 1 \\ -1 \end{pmatrix} = \frac{\hbar}{2\sqrt{2}}\begin{pmatrix} -1 \\ 1 \end{pmatrix}$$

$$= -\frac{\hbar}{2} C_{\vec{\beta}_x},$$

$$M_{\hat{S}_y} C_{\vec{\alpha}_y} = \frac{\hbar}{2\sqrt{2}}\begin{pmatrix} 0 & -i \\ i & 0 \end{pmatrix}\begin{pmatrix} 1 \\ i \end{pmatrix} = \frac{\hbar}{2\sqrt{2}}\begin{pmatrix} 1 \\ i \end{pmatrix}$$

$$= \frac{\hbar}{2} C_{\vec{\alpha}_y},$$

$$M_{\hat{S}_y} C_{\vec{\beta}_y} = \frac{\hbar}{2\sqrt{2}}\begin{pmatrix} 0 & -i \\ i & 0 \end{pmatrix}\begin{pmatrix} 1 \\ -i \end{pmatrix} = \frac{\hbar}{2\sqrt{2}}\begin{pmatrix} -1 \\ i \end{pmatrix}$$

$$= -\frac{\hbar}{2} C_{\vec{\beta}_y}.$$

Q36(13)[7] A unit vector $\vec{n}$ in the 3-dimensional $\vec{\mathbb{E}}^3$ aligned at an angle θ to the z-axis on the xz plane is given by $\vec{n} = \sin\theta\,\vec{i} + \cos\theta\,\vec{k}$. The spin operator $\widehat{S}_{\vec{n}}$ in the direction of the unit vector $\vec{n}$ is given by $\widehat{S}_{\vec{n}} = \vec{n} \cdot \widehat{S}$, where

$$\vec{n} \cdot \widehat{S} = n_x \widehat{S}_x + n_y \widehat{S}_y + n_z \widehat{S}_z = \sin\theta\,\widehat{S}_x + \cos\theta\,\widehat{S}_z.$$

[7] *See* Zettili pp. 298–316 for more examples. $\vec{\mathbb{E}}^3$ corresponds to the physical space we live in.

(a) Show that the matrix representation of $\widehat{S}_{\vec{n}}$ in basis $\{\vec{\alpha}_z, \vec{\beta}_z\}$ is

$$M_{\hat{S}_{\vec{n}}} = \frac{1}{2}\hbar \begin{pmatrix} \cos\theta & \sin\theta \\ \sin\theta & -\cos\theta \end{pmatrix},$$

and that $M_{\hat{S}_{\vec{n}}}$ admits

$$C_{\vec{\eta}\vec{n}+} = \begin{pmatrix} \cos(\theta/2) \\ \sin(\theta/2) \end{pmatrix}, \quad C_{\vec{\eta}\vec{n}-} = \begin{pmatrix} -\sin(\theta/2) \\ \cos(\theta/2) \end{pmatrix} \tag{$*$}$$

as eigenvectors corresponding eigenvalues $\pm\hbar/2$.

(b) Find matrix representation of the state vector of a spin aligned in the direction $\vec{n}$.[8]

(c) A spin is aligned in the positive direction $\vec{n}$. Find the probabilities of a measurement of spin along the z-axis resulting in the values $\pm\frac{1}{2}\hbar$.

(d) A beam of spin-$\frac{1}{2}$ particles with its spin aligned in the positive direction $\vec{n}$ is fed into a Stern–Gerlach apparatus oriented to measure the component of the spin along the z-axis. The incoming beam will split into two with the upper beam corresponding to spin-up along the z-axis and the lower beam corresponding to spin-down along the z-axis. Find the ratio of the intensities of the emerging beams.

SQ36(13)(a) Using the matrix representation $M_{\hat{S}_z}$ of $\widehat{S}_z$ in Eq. (36.143) and the matrix representation $M_{\hat{S}_x}$ of $\widehat{S}_x$ in Eq. (36.144) we immediately obtain the matrix representation $M_{\hat{S}_{\vec{n}}}$ of $\widehat{S}_{\vec{n}}$ in basis $\{\vec{\alpha}, \vec{\beta}\}$, i.e., we have

$$\begin{aligned} M_{\hat{S}_{\vec{n}}} &= (\sin\theta)M_{\hat{S}_x} + (\cos\theta)M_{\hat{S}_z} \\ &= (\sin\theta)\frac{\hbar}{2}\begin{pmatrix} 0 & 1 \\ 1 & 0 \end{pmatrix} + (\cos\theta)\frac{\hbar}{2}\begin{pmatrix} 1 & 0 \\ 0 & -1 \end{pmatrix} \\ &= \frac{1}{2}\hbar \begin{pmatrix} \cos\theta & \sin\theta \\ \sin\theta & -\cos\theta \end{pmatrix}. \end{aligned}$$

[8]The notation $C_{\vec{\eta}\vec{n}}$ in Eq. ($*$) shows that the spin aligned in the direction $\vec{n}$ should be denoted by $\vec{\eta}_{\vec{n}+}$.

Using the formulae

$$\cos(\alpha \pm \beta) = \cos\alpha \cos\beta \mp \sin\alpha \sin\beta,$$

$$\sin(\alpha \pm \beta) = \sin\alpha \cos\beta \pm \cos\alpha \sin\beta,$$

we get

$$\boldsymbol{M}_{\hat{S}_{\vec{n}}} \boldsymbol{C}_{\vec{\eta}_{\vec{n}+}} = \frac{1}{2}\hbar \begin{pmatrix} \cos\theta & \sin\theta \\ \sin\theta & -\cos\theta \end{pmatrix} \begin{pmatrix} \cos(\theta/2) \\ \sin(\theta/2) \end{pmatrix}$$

$$= \frac{1}{2}\hbar \begin{pmatrix} \cos\theta \cos(\theta/2) + \sin\theta \sin(\theta/2) \\ \sin\theta \cos(\theta/2) - \cos\theta \sin(\theta/2) \end{pmatrix}$$

$$= \frac{1}{2}\hbar \begin{pmatrix} \cos(\theta/2) \\ \sin(\theta/2) \end{pmatrix} = \frac{\hbar}{2} \boldsymbol{C}_{\vec{\eta}_{\vec{n}+}},$$

and

$$\boldsymbol{M}_{\hat{S}_{\vec{n}}} \boldsymbol{C}_{\vec{\eta}_{\vec{n}-}} = \frac{1}{2}\hbar \begin{pmatrix} \cos\theta & \sin\theta \\ \sin\theta & -\cos\theta \end{pmatrix} \begin{pmatrix} -\sin(\theta/2) \\ \cos(\theta/2) \end{pmatrix}$$

$$= \frac{1}{2}\hbar \begin{pmatrix} -\cos\theta \sin(\theta/2) + \sin\theta \cos(\theta/2) \\ -\sin\theta \sin(\theta/2) - \cos\theta \cos(\theta/2) \end{pmatrix}$$

$$= \frac{1}{2}\hbar \begin{pmatrix} \sin(\theta/2) \\ -\cos(\theta/2) \end{pmatrix} = -\frac{\hbar}{2} \boldsymbol{C}_{\vec{\eta}_{\vec{n}-}}.$$

These results shows that $\boldsymbol{C}_{\vec{\eta}_{\vec{n}+}}$ and $\boldsymbol{C}_{\vec{\eta}_{\vec{n}-}}$ are eigenvectors of $\boldsymbol{M}_{\hat{S}_{\vec{n}}}$ corresponding to eigenvalues $\hbar/2$ and $-\hbar/2$ respectively.

SQ36(13)(b) The required state vector must be the eigenvector of the spin operator $\widehat{S}_{\vec{n}}$ corresponding to the eigenvalue $\hbar/2$. In matrix representation this means that the matrix representation of the state vector must be the eigenvector of $\boldsymbol{M}_{\hat{S}_{\vec{n}}}$ corresponding to the eigenvalue $\hbar/2$. This eigenvector has been shown in SQ36(13)(a) to be $\boldsymbol{C}_{\vec{\eta}_{\vec{n}+}}$ in Eq. (∗). This explains the notation, i.e., $\vec{\eta}_{\vec{n}+}$ is the state vector for the spin aligned along the $\vec{n}$ direction and $\boldsymbol{C}_{\vec{\eta}_{\vec{n}+}}$ is its matrix representation in basis $\vec{\alpha}_z$, $\vec{\beta}_z$.

SQ36(13)(c) We can express the state vector $\vec{\eta}_{\vec{n}+}$ in terms of a linear combination of the spin-up state vector $\vec{\alpha}_z$ and spin-down state vector $\vec{\beta}_z$. In matrix terms this means that

$$\begin{pmatrix} \cos(\theta/2) \\ \sin(\theta/2) \end{pmatrix} = \cos(\theta/2) \begin{pmatrix} 1 \\ 0 \end{pmatrix} + \sin(\theta/2) \begin{pmatrix} 0 \\ 1 \end{pmatrix}.$$

The required probabilities for spin up and spin down are given by the squares of the coefficients in the above expression, i.e., the probabilities are given respectively by

$$\big(\cos(\theta/2)\big)^2 \quad \text{and} \quad \big(\sin(\theta/2)\big)^2.$$

We can check that total probability is 1, i.e.,

$$\big(\cos(\theta/2)\big)^2 + \big(\sin(\theta/2)\big)^2 = 1.$$

SQ36(13)(d) For the Stern-Gerlach experiment the intensities of the spin-up and spin-down beams are proportional to the probabilities of a particle emerging to be spin-up or spin-down. It follows that the required intensities ratio is

$$\left(\frac{\cos(\theta/2)}{\sin(\theta/2)}\right)^2 = \left(\cot\frac{\theta}{2}\right)^2.$$

Q36(14) The Pauli matrix σ_y is given by

$$\sigma_y = \begin{pmatrix} 0 & -i \\ i & 0 \end{pmatrix}.$$

(a) Show that the Pauli matrix σ_y possesses the following properties[9]:

$$\sigma_y^0 = \sigma_y^2 = \sigma_y^4 = \cdots = \sigma_y^{2k} = \boldsymbol{I}_{2\times 2},$$

$$\sigma_y = \sigma_y^3 = \sigma_y^5 = \cdots = \sigma_y^{2k+1},$$

$$e^{c\sigma_y} = \sum_{k=0}^{\infty} \frac{1}{(2k)!} c^{2k} \boldsymbol{I}_{2\times 2} + \sum_{k=0}^{\infty} \frac{1}{(2k+1)!} c^{2k+1} \sigma_y, \qquad (*)$$

where $\boldsymbol{I}_{2\times 2}$ is the 2×2 identity matrix, $k = 0, 1, 2, 3, \cdots$, and $c \in \mathbb{C}$.

[9] For the exponential function, we can use the expansion

$$e^{c\sigma_y} = \sum_{n=0}^{\infty} \frac{1}{n!} \big(c\sigma_y\big)^n.$$

(b) Show that[10]

$$e^{-i\frac{1}{2}\theta\sigma_y} = \begin{pmatrix} \cos\frac{1}{2}\theta & -\sin\frac{1}{2}\theta \\ \sin\frac{1}{2}\theta & \cos\frac{1}{2}\theta \end{pmatrix}.$$

(c) Let

$$U(\theta) = e^{-\frac{1}{2}i\theta\sigma_y}.$$

Show that $U(\theta)$ is unitary and evaluate the unitary transform of $C_{\vec{a}_z}$ in Eq. (36.139) generated by $U(\theta)$. Give an account on the physical meaning of the transformation.

SQ36(14)(a) First we have σ_y^0 is equal to the 2×2 identity matrix $I_{2\times 2}$ by definition, i.e., by definition the zero power of a square matrix of order n is equal to the identity matrix of the same order. Next we know from Eqs. (7.47) to (7.49) that the square of any of the Pauli matrices are equal to the 2×2 identity matrix, e.g., $\sigma_y^2 = I_{2\times 2}$, results which can be easily verified. It follows that any even power of σ_y is equal to $I_{2\times 2}$. For an odd power of σ_y we have

$$\sigma_y^{2k+1} = \sigma_y^{2k} \cdot \sigma_y = I_{2\times 2} \cdot \sigma_y = \sigma_y.$$

Carrying out the expansion of the exponential in Eq. (∗) we get

$$\begin{aligned} e^{c\sigma_y} &= \sum_{n=0}^{\infty} \frac{1}{n!}\left(c\sigma_y\right)^n \\ &= \sum_{k=0}^{\infty} \frac{1}{(2k)!}\left(c\sigma_y\right)^{2k} + \sum_{k=0}^{\infty} \frac{1}{(2k+1)!}\left(c\sigma_y\right)^{(2k+1)} \\ &= \sum_{k=0}^{\infty} \frac{1}{(2k)!}\, c^{2k}\, I_{2\times 2} + \sum_{k=0}^{\infty} \frac{1}{(2k+1)!}\, c^{(2k+1)}\, \sigma_y. \end{aligned}$$

[10]Use the following expansions of $\cos x$ and $\sin x$:

$$\cos x = \sum_{k=0}^{\infty} \frac{(-1)^k}{(2k)!}(x)^{2k}, \quad \sin x = \sum_{k=0}^{\infty} \frac{(-1)^k}{(2k+1)!}(x)^{2k+1}.$$

SQ36(14)(b) Replacing c by $-i\frac{1}{2}\theta$ in the above expansion of the exponential we obtain an expansion of $\exp -i\frac{1}{2}\theta\sigma_y$, i.e.,

$$\exp\Big(-i\frac{1}{2}\theta\sigma_y\Big) = \sum_{k=0}^{\infty} \frac{1}{(2k)!}\,(-i\theta/2)^{2k}\,\boldsymbol{I}_{2\times 2}$$

$$= +\sum_{k=0}^{\infty} \frac{1}{(2k+1)!}\,(-i(\theta/2))^{(2k+1)}\,\sigma_y.$$

Using the formular $x^{mn} = (x^n)^m$ we get

$$(-i)^{2k} = \big((-i)^2\big)^k = (-1)^k,$$

$$(-i)^{(2k+1)} = (-i)^{2k}\,(-i) = (-1)^k(-i).$$

The exponential function $\exp -i\frac{1}{2}\theta\sigma_y$ becomes

$$\exp\Big(-i\frac{1}{2}\theta\sigma_y\Big) = \sum_{k=0}^{\infty} \frac{(-1)^k}{(2k)!}\,(\theta/2)^{2k}\,\boldsymbol{I}_{2\times 2}$$

$$-i\sum_{k=0}^{\infty} \frac{(-1)^k}{(2k+1)!}\,(\theta/2)^{(2k+1)}\,\sigma_y$$

$$= \cos(\theta/2)\,\boldsymbol{I}_{2\times 2} - i\sin(\theta/2)\,\sigma_y$$

$$= \cos(\theta/2)\begin{pmatrix}1 & 0\\ 0 & 1\end{pmatrix} - i\sin(\theta/2)\begin{pmatrix}0 & -i\\ i & 0\end{pmatrix}$$

$$= \cos(\theta/2)\begin{pmatrix}1 & 0\\ 0 & 1\end{pmatrix} + \sin\theta/2\begin{pmatrix}0 & -1\\ 1 & 0\end{pmatrix},$$

$$= \begin{pmatrix}\cos(\theta/2) & -\sin(\theta/2)\\ \sin(\theta/2) & \cos(\theta/2)\end{pmatrix}.$$

SQ36(14)(c) A square matrix is unitary if it satisfies Eq. (7.158), i.e., its adjoint is equal to its inverse. The adjoint of $\boldsymbol{U}(\theta)$ is

$$\begin{pmatrix}\cos(\theta/2) & \sin(\theta/2)\\ -\sin(\theta/2) & \cos(\theta/2)\end{pmatrix}.$$

This is equal to the inverse of $\boldsymbol{U}(\theta)$ since

$$\begin{pmatrix}\cos(\theta/2) & \sin(\theta/2)\\ -\sin(\theta/2) & \cos(\theta/2)\end{pmatrix}\begin{pmatrix}\cos(\theta/2) & -\sin(\theta/2)\\ \sin(\theta/2) & \cos(\theta/2)\end{pmatrix} = \begin{pmatrix}1 & 0\\ 0 & 1\end{pmatrix}.$$

The unitary transform of $\boldsymbol{C}_{\vec{\alpha}_z}$ generated by $\boldsymbol{U}(\theta)$ is given by

$$\boldsymbol{U}(\theta)\boldsymbol{C}_{\vec{\alpha}_z} = \begin{pmatrix} \cos(\theta/2) & -\sin(\theta/2) \\ \sin(\theta/2) & \cos(\theta/2) \end{pmatrix} \begin{pmatrix} 1 \\ 0 \end{pmatrix}$$

$$= \begin{pmatrix} \cos(\theta/2) \\ \sin(\theta/2) \end{pmatrix} = \boldsymbol{C}_{\vec{\eta}_{\vec{n}+}}.$$

Since $\boldsymbol{C}_{\vec{\alpha}_z}$ in Eq. (36.139) and $\boldsymbol{C}_{\vec{\eta}_{\vec{n}+}}$ in Eq. (∗) correspond to the spin orientation along the positive directions of z-axis and the vector $\vec{n}$ in Q36(13) respectively the transformation from $\boldsymbol{C}_{\vec{\alpha}_z}$ to $\boldsymbol{U}(\theta)\boldsymbol{C}_{\vec{\alpha}_z}$ corresponds to a change of spin orientation from z-axis to $\vec{n}$.

Chapter 37

Particles in Static Magnetic Fields

Q37(1) Verify Eq. (37.2).

SQ37(1) The vector potential is

$$\vec{A} = -\frac{1}{2} yB\,\vec{i} + \frac{1}{2} xB\,\vec{j}.$$

Using the expression for the curl in Eq. (27.33) we get, due to $A_z = 0$,

$$\nabla \times \vec{A} = \left(-\frac{\partial A_y}{\partial z}\right)\vec{i} + \left(\frac{\partial A_x}{\partial z}\right)\vec{j} + \left(\frac{\partial A_y}{\partial x} - \frac{\partial A_x}{\partial y}\right)\vec{k}$$

$$= B\,\vec{k}.$$

Q37(2) Using Eqs. (37.8) to (37.10), show that the magnetic field in Eq. (37.7) is derivable from the vector potential in Eq. (37.14) and that the magnetic field in Eq. (37.16) is derivable from the vector potential in Eq. (37.17).

SQ37(2) In cylindrical coordinates the vector potential $\vec{A}$ in Eq. (37.14) has a non-zero component only in the θ direction, i.e.,

$$A_r = 0, \quad A_z = 0 \quad \text{and} \quad A_\theta(r) = \frac{1}{2} Br.$$

Then Eqs. (37.8), (37.9) and (37.10) tell us that:

(1) Since r and A_θ are independent of z and $A_z = 0$ we get $B_r = 0$.

Quantum Mechanics: Problems and Solutions
K. Kong Wan

ISBN 978-981-4800-72-3 (Paperback), 978-0-429-29647-5 (eBook)
www.jennystanford.com

(2) Since $A_z = 0$ and $A_r = 0$ have $B_\theta = 0$.
(3) So, the curl, and hence the resulting magnetic field, has only an z component $B_z(r)$ given by

$$B_z = \frac{1}{r}\left(\frac{\partial(rA_\theta)}{\partial r} - \frac{\partial A_r}{\partial\theta}\right) = \frac{1}{r}\frac{\partial(Br^2/2)}{\partial r} = B.$$

For the vector potential in Eq. (37.17) we have $A_r = 0,\ \ A_z = 0$ and

$$A_\theta(r) = \begin{cases} Br/2, & r < b \text{ (proportional to } r) \\ \Phi_b/2\pi r, & r > b \text{ (inversely proportional to } r) \end{cases},$$

where $\Phi_b = \pi b^2 B$ is the magnetic flux enclosed by the circle of radius b in the x-y plane centered at the origin. To show that this is the correct potential for the given field we can carry out the following calculations:

(1) Since $A_z = 0, r$ and A_θ are independent of z we have $B_r = 0$.
(2) Since $A_z = 0$ and $A_r = 0$ have $B_\theta = 0$.
(3) So, the curl, and hence the resulting magnetic field, has only a z component $B_z(r)$ given by

$$B_z = \frac{1}{r}\left(\frac{\partial(rA_\theta)}{\partial r} - \frac{\partial A_r}{\partial\theta}\right) = \begin{cases} \dfrac{1}{r}\dfrac{\partial(Br^2/2)}{\partial r} = B & \text{for } \ r < R \\ \dfrac{1}{r}\left(\dfrac{\partial(\Phi_b/2\pi)}{\partial r}\right) = 0 & \text{for } \ r > R \end{cases}.$$

Q37(3) When spatial motion is neglected, the Hamiltonian of an electron of charge $-e$ and mass m in a uniform and static magnetic field of magnitude B pointing along the z-direction is given by

$$\widehat{H}^{(s)} = \frac{e}{m}B\widehat{S}_z.$$

Write down the Pauli–Schrödinger equation for the evolution of a spin state in the Schrödinger picture. Show that the following initial spin state

$$\vec{\eta}(0) = \frac{1}{\sqrt{2}}\left\{\vec{\alpha}_z + \vec{\beta}_z\right\}$$

will evolve to a new state $\vec{\eta}(t)$ at time t given

$$\vec{\eta}(t) = \frac{1}{\sqrt{2}}\left\{e^{-\frac{1}{2}i\omega t}\vec{\alpha}_z + e^{\frac{1}{2}i\omega t}\vec{\beta}_z\right\},$$

where $\omega = eB/m$. What are the spin orientations initially at $t = 0$ and later at $t = \pi/2\omega$?

SQ37(3) The Pauli-Schrödinger equation is

$$i\hbar\frac{d}{dt}\vec{\eta}(t) = \hat{H}^{(s)}\vec{\eta}(t) = \frac{eB}{m}\hat{S}_z\vec{\eta}(t).$$

The given vector $\vec{\eta}(t)$ satisfies the initial condition, i.e., we have

$$\vec{\eta}(0) = \vec{\eta}(t=0) = \frac{1}{\sqrt{2}}(\vec{\alpha}_z + \vec{\beta}_z).$$

The given vector $\vec{\eta}(t)$ also satisfies the Pauli-Schrödinger equation, i.e., we have

$$\begin{aligned}
i\hbar\frac{d\vec{\eta}(t)}{dt} &= i\hbar\frac{1}{\sqrt{2}}\frac{d}{dt}\left\{e^{-\frac{1}{2}i\omega t}\vec{\alpha}_z + e^{\frac{1}{2}i\omega t}\vec{\beta}_z\right\} \\
&= i\hbar\frac{1}{\sqrt{2}}\left\{-\frac{1}{2}i\omega e^{-\frac{1}{2}i\omega t}\vec{\alpha}_z + \frac{1}{2}i\omega e^{\frac{1}{2}i\omega t}\vec{\beta}_z\right\} \\
&= \frac{1}{2}\hbar\omega\frac{1}{\sqrt{2}}\left\{e^{-\frac{1}{2}i\omega t}\vec{\alpha}_z - e^{\frac{1}{2}i\omega t}\vec{\beta}_z\right\}.
\end{aligned}$$

$$\begin{aligned}
\hat{H}\vec{\eta}(t) &= \frac{eB}{m}\frac{1}{\sqrt{2}}\left\{e^{-\frac{1}{2}i\omega t}\hat{S}_z\vec{\alpha}_z + e^{\frac{1}{2}i\omega t}\hat{S}_z\vec{\beta}_z\right\} \\
&= \frac{eB\cdot 1}{m\sqrt{2}}\left\{e^{-\frac{1}{2}i\omega t}\frac{\hbar}{2}\vec{\alpha}_z + e^{\frac{1}{2}i\omega t}\frac{-\hbar}{2}\vec{\beta}_z\right\} \\
&= \frac{\hbar}{2}\frac{eB}{m}\frac{1}{\sqrt{2}}\left\{e^{-\frac{1}{2}i\omega t}\vec{\alpha}_z - e^{\frac{1}{2}i\omega t}\vec{\beta}_z\right\} \\
&= i\hbar\frac{d\vec{\eta}(t)}{dt} \quad \text{since } \frac{eB}{m} = \omega.
\end{aligned}$$

Hence, $\vec{\eta}(t)$ is a solution of the Pauli-Schrödinger equation. We can conclude that $\vec{\eta}(t)$ is the state vector evolved from the initial state vector $\vec{\eta}(0)$.

According to Eq. (36.137) the initial state vector $\vec{\eta}(0)$ is an eigenvector of $\hat{S}_x$ corresponding to eigenvalue $\hbar/2$. This implies that initially the spin is oriented in the positive x-direction. At time

$t = \pi/2\omega$ we have

$$\begin{aligned}
\vec{\eta}(\pi/2\omega) &= \frac{1}{\sqrt{2}}\left\{e^{-\frac{1}{2}i\omega\frac{\pi}{2\omega}}\vec{\alpha}_z + e^{\frac{1}{2}i\omega\frac{\pi}{2\omega}}\vec{\beta}_z\right\} \\
&= \frac{1}{\sqrt{2}}\left\{e^{-i\frac{\pi}{4}}\vec{\alpha}_z + e^{i\frac{\pi}{4}}\vec{\beta}_z\right\} \\
&= \frac{1}{\sqrt{2}}e^{-i\frac{\pi}{4}}\left\{\vec{\alpha}_z + e^{i\frac{\pi}{4}}e^{i\frac{\pi}{4}}\vec{\beta}_z\right\} \\
&= \frac{1}{\sqrt{2}}e^{-i\frac{\pi}{4}}\left(\vec{\alpha}_z + e^{i\frac{\pi}{2}}\vec{\beta}_z\right) \\
&= \frac{1}{\sqrt{2}}e^{-i\frac{\pi}{4}}\left(\vec{\alpha}_z + i\,\vec{\beta}_z\right).
\end{aligned}$$

According Eq. (36.138) this represents a spin orientation along positive y-direction. The complex phase factor does not affect this conclusion. The spin evolves from x-direction to y-direction.

Bibliography

Fano, G. (1992). *Mathematical Methods of Quantum Mechanics*, McGraw-Hill, New York.

Halmos, P. R. (1958). *Finite-Dimensional Vector Spaces*, Van Nostrand, Princeton.

Isham, C. J. (1995). *Lectures on Quantum Theory*, Imperial College Press, London.

Jauch, J. M. (1968). *Foundations of Quantum Mechanics*, Addison-Wesley, Reading, Mass.

Jordan, T. F. (1969). *Linear Operators for Quantum Mechanics*, Thomas F. Jordan, Duluth, Minnesota, USA.

Merzbacher, E. (1998). *Quantum Mechanics*, 3rd edition, Wiley, New York.

Papoulis, A. (1965). *Probability, Random Variables, and Stochastic Processes*, MaGraw-Hill, New York.

Prugovečki, E. (1981). *Quantum Mechanics in Hilbert Space*, 2nd edition, Academic Press, New York.

Roman, P. (1975). *Some Modern Mathematics for Physicists and Other Outsiders*, Vol. 2, Pergamon, New York.

Wan, K. K. (2006). *From Micro to Macro Quantum Systems*, Imperial College Press, London.

Wan, K. K. (2019). *Quantum Mechanics: A Fundamental Approach*, Jenny Stanford Publishing, Singapore.

Weidmann, J. (1980). *Linear Operators in Hilbert Spaces*, Springer-Verlag, New York.

Zettili, N. (2001). *Quantum Mechanics*, Wiley, Chichester.

Printed in the United States
By Bookmasters

Printed in the United States
By Bookmasters